A Course in Financial Calculus

A Course in
Financial Calculus

Alison Etheridge

University of Oxford

PUBLISHED BY THE PRESS SYNDICATE OF THE UNIVERSITY OF CAMBRIDGE
The Pitt Building, Trumpington Street, Cambridge, United Kingdom

CAMBRIDGE UNIVERSITY PRESS
The Edinburgh Building, Cambridge CB2 2RU, UK
40 West 20th Street, New York, NY 10011-4211, USA
477 Williamstown Road, Port Melbourne, VIC 3207, Australia
Ruiz de Alarcón 13, 28014 Madrid, Spain
Dock House, The Waterfront, Cape Town 8001, South Africa

http://www.cambridge.org

First published 2002, reprinted with corrections 2004 (twice)

Printed in the United Kingdom at the University Press, Cambridge

Typeface Times 10/13pt.   *System* LATEX 2_ε [DBD]

A catalogue record of this book is available from the British Library

ISBN 0 521 81385 9 hardback
ISBN 0 521 89077 2 paperback

Contents

Preface

Financial mathematics provides a striking example of successful collaboration between academia and industry. Advanced mathematical techniques, developed in both universities and banks, have transformed the derivatives business into a multi-trillion-dollar market. This has led to demand for highly trained students and with that demand comes a need for textbooks.

This volume provides a first course in financial mathematics. The influence of *Financial Calculus* by Martin Baxter and Andrew Rennie will be obvious. I am extremely grateful to Martin and Andrew for their guidance and for allowing me to use some of the material from their book.

The structure of the text largely follows *Financial Calculus*, but the mathematics, especially the discussion of stochastic calculus, has been expanded to a level appropriate to a university mathematics course and the text is supplemented by a large number of exercises. In order to keep the course to a reasonable length, some sacrifices have been made. Most notable is that there was not space to discuss interest rate models, although many of the most popular ones do appear as examples in the exercises. As partial compensation, the necessary mathematical background for a rigorous study of interest rate models is included in Chapter 7, where we briefly discuss some of the topics that one might hope to include in a *second* course in financial mathematics. The exercises should be regarded as an integral part of the course. Solutions to these are available to *bona fide* teachers from solutions@cambridge.org.

The emphasis is on stochastic techniques, but not to the exclusion of all other approaches. In common with practically every other book in the area, we use binomial trees to introduce the ideas of arbitrage pricing. Following *Financial Calculus*, we also present discrete versions of key definitions and results on martingales and stochastic calculus in this simple framework, where the important ideas are not obscured by analytic technicalities. This paves the way for the more technical results of later chapters. The connection with the partial differential equation approach to arbitrage pricing is made through both delta-hedging arguments and the Feynman–Kac Stochastic Representation Theorem. Whatever approach one adopts, the key point that we wish to emphasise is that since the theory rests on the assumption of absence of arbitrage, hedging is vital. Our pricing formulae only make sense if there is a 'replicating portfolio'.

An early version of this course was originally delivered to final year undergraduate and first year graduate mathematics students in Oxford in 1997/8. Although we assumed some familiarity with probability theory, this was not regarded as a prerequisite and students on those courses had little difficulty picking up the necessary concepts as we met them. Some suggestions for suitable background reading are made in the bibliography. Since a first course can do little more than scratch the surface of the subject, we also make suggestions for supplementary and more advanced reading from the bewildering array of available books.

This project was supported by an EPSRC Advanced Fellowship. It is a pleasure and a privilege to work in Magdalen College and my thanks go to the President, Fellows, staff and students for making it such an exceptional environment. Many people have made helpful suggestions or read early drafts of this volume. I should especially like to thank Ben Hambly, Alex Jackson and Saurav Sen. Thanks also to David Tranah at CUP who played a vital rôle in shaping the project. His input has been invaluable. Most of all, I should like to thank Lionel Mason for his constant support and encouragement.

Alison Etheridge, June 2001

1 Single period models

Summary

In this chapter we introduce some basic definitions from finance and investigate the problem of pricing financial instruments in the context of a very crude model. We suppose the market to be observed at just two times: zero, when we enter into a financial contract; and T, the time at which the contract expires. We further suppose that the market can only be in one of a finite number of states at time T. Although simplistic, this model reveals the importance of the central paradigm of modern finance: the idea of a perfect hedge. It is also adequate for a preliminary discussion of the notion of 'complete market' and its importance if we are to find a 'fair' price for our financial contract.

The proofs in §1.5 can safely be omitted, although we shall from time to time refer back to the statements of the results.

1.1 Some definitions from finance

Financial market instruments can be divided into two types. There are the *underlying* stocks – shares, bonds, commodities, foreign currencies; and their *derivatives*, claims that promise some payment or delivery in the future contingent on an underlying stock's behaviour. Derivatives can reduce risk – by enabling a player to fix a price for a future transaction now – or they can magnify it. A costless contract agreeing to pay off the difference between a stock and some agreed future price lets both sides ride the risk inherent in owning a stock, without needing the capital to buy it outright.

The connection between the two types of instrument is sufficiently complex and uncertain that both trade fiercely in the same market. The apparently random nature of the underlying stocks filters through to the derivatives – they appear random too.

Derivatives   Our central purpose is to determine how much one should be willing to pay for a derivative security. But first we need to learn a little more of the language of finance.

Definition 1.1.1   A forward contract *is an agreement to buy (or sell) an asset on a specified future date, T, for a specified price, K. The buyer is said to hold the* long *position, the seller the* short *position.*

Forwards are not generally traded on exchanges. It costs nothing to enter into a forward contract. The 'pricing problem' for a forward is to determine what value of K should be written into the contract. A *futures contract* is the same as a forward except that futures *are* normally traded on exchanges and the exchange specifies certain standard features of the contract and a particular form of settlement.

Forwards provide the simplest examples of derivative securities and the mathematics of the corresponding pricing problem will also be simple. A much richer theory surrounds the pricing of *options*. An option gives the holder the *right*, but not the *obligation*, to do something. Options come in many different guises. Black and Scholes gained fame for pricing a European call option.

Definition 1.1.2   A European call option *gives the holder the right, but not the obligation, to buy an asset at a specified time, T, for a specified price, K.*
      A European put option *gives the holder the right to* sell *an asset for a specified price, K, at time T.*

In general *call* refers to buying and *put* to selling. The term *European* is reserved for options whose value to the holder at the time, T, when the contract expires depends on the state of the market only at time T. There are other options, for example American options or Asian options, whose payoff is contingent on the behaviour of the underlying over the whole time interval $[0, T]$, but the technology of this chapter will only allow meaningful discussion of European options.

Definition 1.1.3   *The time, T, at which the derivative contract expires is called the* exercise date *or the* maturity. *The price K is called the* strike price.

The pricing problem

So what is the pricing problem for a European call option? Suppose that a company has to deal habitually in an intrinsically risky asset such as oil. They may for example know that in three months time they will need a thousand barrels of crude oil. Oil prices can fluctuate wildly, but by purchasing European call options, with strike K say, the company knows the *maximum* amount of money that it will need (in three months time) in order to buy a thousand barrels. One can think of the option as insurance against increasing oil prices. The pricing problem is now to determine, for given T and K, how much the company should be willing to pay for such insurance.

For this example there is an extra complication: it costs money to store oil. To simplify our task we are first going to price derivatives based on assets that can be held without additional cost, typically company shares. Equally we suppose that there is no additional benefit to holding the shares, that is no dividends are paid.

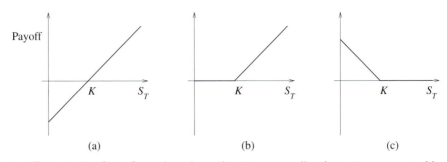

Figure 1.1   Payoff at maturity of (a) a forward purchase, (b) a European call and (c) a European put with strike K as a function of S_T.

> **Assumption**   Unless otherwise stated, the underlying asset can be held without additional cost or benefit.

This assumption will be relaxed in Chapter 5.

Suppose then that our company enters into a contract that gives them the right, but not the obligation, to buy one unit of stock for price K in three months time. How much should they pay for this contract?

Payoffs

As a first step, we need to know what the contract will be worth at the expiry date. If at the time when the option expires (three months hence) the actual price of the underlying stock is S_T and $S_T > K$ then the option will be exercised. The option is then said to be *in the money*: an asset worth S_T can be purchased for just K. The value to the company of the option is then $(S_T - K)$. If, on the other hand, $S_T < K$, then it will be cheaper to buy the underlying stock on the open market and so the option will not be exercised. (It is this freedom *not* to exercise that distinguishes options from futures.) The option is then worthless and is said to be *out of the money*. (If $S_T = K$ the option is said to be *at the money*.) The *payoff* of the option at time T is thus

$$(S_T - K)_+ \triangleq \max\{(S_T - K), 0\}.$$

Figure 1.1 shows the payoff at maturity of three derivative securities: a forward purchase, a European call and a European put, each as a function of stock price at maturity. Before embarking on the valuation *at time zero* of derivative contracts, we allow ourselves a short aside.

Packages

We have presented the European call option as a means of reducing risk. Of course it can also be used by a speculator as a bet on an increase in the stock price. In fact by holding *packages*, that is combinations of the 'vanilla' options that we have described so far, we can take rather complicated bets. We present just one example; more can be found in Exercise 1.

Example 1.1.4 (A straddle)   *Suppose that a speculator is expecting a large move in a stock price, but does not know in which direction that move will be. Then a possible combination is a* straddle. *This involves holding a European call and a European put with the same strike price and maturity.*

Explanation: The payoff of this straddle is $(S_T - K)_+$ (from the call) plus $(K - S_T)_+$ (from the put), that is, $|S_T - K|$. Although the payoff of this combination is always positive, if, at the expiry time, the stock price is too close to the strike price then the payoff will not be sufficient to offset the cost of purchasing the options and the investor makes a loss. On the other hand, large movements in price can lead to substantial profits.   □

1.2   Pricing a forward

In order to solve our pricing problems, we are going to have to make some assumptions about the way in which markets operate. To formulate these we begin by discussing forward contracts in more detail.

Recall that a forward contract is an agreement to buy (or sell) an asset on a specified future date for a specified price. Suppose then that I agree to buy an asset for price K at time T. The payoff at time T is just $S_T - K$, where S_T is the actual asset price at time T. The payoff could be positive or it could be negative and, since the cost of entering into a forward contract is zero, this is also my total gain (or loss) from the contract. Our problem is to determine the fair value of K.

Expectation pricing   At the time when the contract is written, we don't know S_T, we can only guess at it, or, more formally, assign a probability distribution to it. A widely used model (which underlies the Black–Scholes analysis of Chapter 5) is that stock prices are *lognormally distributed*. That is, there are constants ν and σ such that the *logarithm* of S_T/S_0 (the stock price at time T divided by that at time zero, usually called the *return*) is normally distributed with mean ν and variance σ^2. In symbols:

$$\mathbb{P}\left[\frac{S_T}{S_0} \in [a, b]\right] = \mathbb{P}\left[\log\left(\frac{S_T}{S_0}\right) \in [\log a, \log b]\right]$$
$$= \int_{\log a}^{\log b} \frac{1}{\sqrt{2\pi}\sigma} \exp\left(-\frac{(x - \nu)^2}{2\sigma^2}\right) dx.$$

Notice that stock prices, and therefore a and b, should be positive, so that the integral on the right hand side is well defined.

Our first guess might be that $\mathbb{E}[S_T]$ should represent a fair price to write into our contract. However, it would be a rare coincidence for this to be the market price. In fact we'll show that the cost of borrowing is the key to our pricing problem.

The risk-free rate   We need a model for the *time value of money*: a dollar now is worth more than a dollar promised at some later time. We assume a market for these future promises (the *bond* market) in which prices are derivable from some interest rate. Specifically:

> **Time value of money**   We assume that for any time T less than some horizon τ the value now of a dollar promised at T is e^{-rT} for some constant $r > 0$. The rate r is then the *continuously compounded* interest rate for this period.

Such a market, derived from say US Government bonds, carries no risk of default – the promise of a future dollar will always be honoured. To emphasise this we will often refer to r as the *risk-free interest rate*. In this model, by buying or selling cash bonds, investors can borrow money for the same risk-free rate of interest as they can lend money.

Interest rate markets are not this simple in practice, but that is an issue that we shall defer.

Arbitrage pricing

We now show that it is the *risk-free interest rate*, or equivalently the price of a cash bond, and not our lognormal model that forces the choice of the strike price, K, upon us in our forward contract.

Interest rates will be different for different currencies and so, for definiteness, suppose that we are operating in the dollar market, where the (risk-free) interest rate is r.

- Suppose first that $K > S_0 e^{rT}$. The seller, obliged to deliver a unit of stock for $\$K$ at time T, adopts the following strategy: she borrows $\$S_0$ at time zero (i.e. sells bonds to the value $\$S_0$) and buys one unit of stock. At time T, she must repay $\$S_0 e^{rT}$, but she has the stock to sell for $\$K$, leaving her a *certain* profit of $\$(K - S_0 e^{rT})$.
- If $K < S_0 e^{rT}$, then the buyer reverses the strategy. She sells a unit of stock at time zero for $\$S_0$ and buys cash bonds. At time T, the bonds deliver $\$S_0 e^{rT}$ of which she uses $\$K$ to buy back a unit of stock leaving her with a *certain* profit of $\$(S_0 e^{rT} - K)$.

Unless $K = S_0 e^{rT}$, one party is guaranteed to make a profit.

Definition 1.2.1   *An opportunity to lock into a risk-free profit is called an* arbitrage opportunity.

The starting point in establishing a model in modern finance theory is to specify that there is no arbitrage. (In fact there are people who make their living entirely from exploiting arbitrage opportunities, but such opportunities do not exist for a significant length of time before market prices move to eliminate them.) We have proved the following lemma.

Lemma 1.2.2   *In the absence of arbitrage, the strike price in a forward contract with expiry date T on a stock whose value at time zero is S_0 is $K = S_0 e^{rT}$, where r is the risk-free rate of interest.*

The price $S_0 e^{rT}$ is sometimes called the *arbitrage price*. It is also known as the *forward price* of the stock.

6

SINGLE PERIOD MODELS

Remark: In our proof of Lemma 1.2.2, the buyer sold stock that she may not own. This is known as *short selling*. This can, and does, happen: investors can 'borrow' stock as well as money. □

Of course forwards are a very special sort of derivative. The argument above won't tell us how to value an option, but the strategy of seeking a price that does not provide either party with a risk-free profit will be fundamental in what follows.

Let us recap what we have done. In order to price the forward, we constructed a portfolio, comprising one unit of underlying stock and $-S_0$ cash bonds, whose value at the maturity time T is *exactly* that of the forward contract itself. Such a portfolio is said to be a *perfect hedge* or *replicating portfolio*. This idea is the central paradigm of modern mathematical finance and will recur again and again in what follows. Ironically we shall use expectation repeatedly, but as a tool in the construction of a perfect hedge.

1.3 The one-step binary model

We are now going to turn to establishing the fair price for European call options, but in order to do so we first move to a *simpler* model for the movement of market prices. Once again we suppose that the market is observed at just two times, that at which the contract is struck and the expiry date of the contract. Now, however, we shall suppose that there are just two possible values for the stock price at time T. We begin with a simple example.

Pricing a European call

Example 1.3.1 *Suppose that the current price in Japanese Yen of a certain stock is ¥2500. A European call option, maturing in six months time, has strike price ¥3000. An investor believes that with probability one half the stock price in six months time will be ¥4000 and with probability one half it will be ¥2000. He therefore calculates the expected value of the option (when it expires) to be ¥500. The riskless borrowing rate in Japan is currently zero and so he agrees to pay ¥500 for the option. Is this a fair price?*

Solution: In the light of the previous section, the reader will probably have guessed that the answer to this question is no. Once again, we show that one party to this contract can make a risk-free profit. In this case it is the seller of the contract. Here is just one of the many possible strategies that she could adopt.

Strategy: At time zero, sell the option, borrow ¥2000 and buy a unit of stock.

- Suppose first that at expiry the price of the stock is ¥4000; then the contract will be exercised and so she must sell her stock for ¥3000. She then holds ¥(−2000+3000). That is ¥1000.
- If, on the other hand, at expiry the price of the stock is ¥2000, then the option will not be exercised and so she sells her stock on the open market for just ¥2000. Her

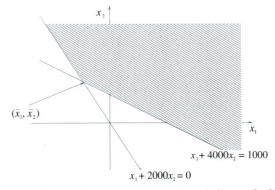

Figure 1.2   The seller of the contract in Example 1.3.1 is guaranteed a risk-free profit if she can buy any portfolio in the shaded region.

net cash holding is then $¥(-2000 + 2000)$. That is, she exactly breaks even.

Either way, our seller has a positive chance of making a profit with *no risk* of making a loss. The price of the option is too high.

So what is the right price for the option?

Let's think of things from the point of view of the seller. Writing S_T for the price of the stock when the contract expires, she knows that at time T she needs $¥(S_T - 3000)_+$ in order to meet the claim against her. The idea is to calculate how much money she needs at time zero, to be held in a combination of stocks and cash, to guarantee this.

Suppose then that she uses the money that she receives for the option to buy a portfolio comprising x_1 Yen and x_2 stocks. If the price of the stock is $¥4000$ at expiry, then the time T value of the portfolio is $x_1 e^{rT} + 4000x_2$. The seller of the option requires this to be at least $¥1000$. That is, since interest rates are zero,

$$x_1 + 4000x_2 \geq 1000.$$

If the price is $¥2000$ she just requires the value of the portfolio to be non-negative,

$$x_1 + 2000x_2 \geq 0.$$

A profit is guaranteed (without risk) for the seller if (x_1, x_2) lies in the interior of the shaded region in Figure 1.2. On the boundary of the region, there is a positive probability of profit and no probability of loss at all points other than the intersection of the two lines. The portfolio represented by the point $(\bar{x}_1, \bar{x}_2)$ will provide *exactly* the wealth required to meet the claim against her at time T.

Solving the simultaneous equations gives that the seller can exactly meet the claim if $\bar{x}_1 = -1000$ and $\bar{x}_2 = 1/2$. The cost of building this portfolio at time zero is $¥(-1000 + 2500/2)$, that is $¥250$. For any price higher than $¥250$, the seller can make a risk-free profit.

If the option price is *less* than ¥250, then the *buyer* can make a risk-free profit by 'borrowing' the portfolio $(\overline{x}_1, \overline{x}_2)$ and buying the option. In the absence of arbitrage then, the fair price for the option is ¥250.   □

Notice that just as for our forward contract, we did not use the probabilities that we assigned to the possible market movements to arrive at the fair price. We just needed the fact that we could *replicate* the claim by this simple portfolio. The seller can *hedge* the *contingent claim* ¥$(S_T - 3000)_+$ using the *portfolio* consisting of ¥x_1 and x_2 units of stock.

Pricing formula for European options

One can use exactly the same argument to prove the following result.

Lemma 1.3.2   *Suppose that the risk-free dollar interest rate (to a time horizon $\tau > T$) is r. Denote the time zero (dollar) value of a certain asset by S_0. Suppose that the motion of stock prices is such that the value of the asset at time T will be either S_0u or S_0d. Assume further that*

$$d < e^{rT} < u.$$

At time zero, the market price of a European option with payoff $C(S_T)$ at the maturity T is

$$\left(\frac{1 - de^{-rT}}{u - d}\right) C(S_0u) + \left(\frac{ue^{-rT} - 1}{u - d}\right) C(S_0d).$$

Moreover, the seller of the option can construct a portfolio whose value at time T is exactly $(S_T - K)_+$ by using the money received for the option to buy

$$\phi \triangleq \frac{C(S_0u) - C(S_0d)}{S_0u - S_0d} \tag{1.1}$$

units of stock at time zero and holding the remainder in bonds.

The proof is Exercise 4(a).

1.4   A ternary model

There were several things about the binary model that were very special. In particular we assumed that we knew that the asset price would be one of just two specified values at time T. What if we allow *three* values?

We can try to repeat the analysis of §1.3. Again the seller would like to replicate the claim at time T by a portfolio consisting of ¥x_1 and x_2 stocks. This time there will be three scenarios to consider, corresponding to the three possible values of S_T. If interest rates are zero, this gives rise to the three inequalities

$$x_1 + S_T^i x_2 \geq (S_T^i - 3000)_+, \qquad i = 1, 2, 3,$$

where S_T^i are the possible values of S_T. The picture is now something like that in Figure 1.3.

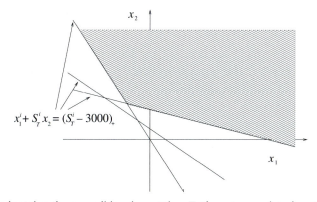

$$x_1^i + S_T^i x_2 = (S_T^i - 3000)_+$$

Figure 1.3   If the stock price takes three possible values at time T, then at any point where the seller of the option has no risk of making a loss, she has a *strictly positive* chance of making a profit.

In order to be *guaranteed* to meet the claim at time T, the seller requires (x_1, x_2) to lie in the shaded region, but at any point in that region, she has a strictly positive probability of making a profit and zero probability of making a loss. Any portfolio from outside the shaded region carries a risk of a loss. There is no portfolio that *exactly* replicates the claim and there is no unique 'fair' price for the option.

Our market is not *complete*. That is, there are contingent claims that cannot be perfectly hedged.

Bigger models

Of course we are tying our hands in our efforts to hedge a claim. First, we are only allowing ourselves portfolios consisting of the underlying stock and cash bonds. Real markets are bigger than this. If we allow ourselves to trade in a third 'independent' asset, then our analysis leads to three non-parallel planes in $\mathbb{R}^3$. These *will* intersect in a single point representing a portfolio that exactly replicates the claim. This then raises a question: when is there arbitrage in larger market models? We shall answer this question for a single period model in the next section. The second constraint that we have placed upon ourselves is that we are not allowed to adjust our portfolio between the time of the selling of the contract and its maturity. In fact, as we see in Chapter 2, if we consider the market to be observable at intermediate times between zero and T, and allow our seller to rebalance her portfolio at such times (without changing its value), then we *can* allow any number of possible values for the stock price at time T and yet still replicate each claim at time T by a portfolio consisting of just the underlying and cash bonds.

1.5 A characterisation of no arbitrage

In our binary setting it was easy to find the right price for an option simply by solving a pair of simultaneous equations. However, the binary model is very special and, after our experience with the ternary model, alarm bells may be ringing. The binary model describes the evolution of just one stock (and one bond). One solution to our difficulties with the ternary model was to allow trade in another 'independent' asset. In this section we extend this idea to larger market models and characterise those models for which there are a sufficient number of independent assets that any option has a fair price. Other than Definition 1.5.1 and the statement of Theorem 1.5.2, this section can safely be omitted.

A market with N assets

Our market will now consist of a finite (but possibly large) number of tradable assets. Again we restrict ourselves to single period models, in which the market is observable only at time zero and a fixed future time T. However, the extension to multiple time periods exactly mirrors that for binary models that we describe in §2.1.

Suppose then that there are N tradable assets in the market. Their prices at time zero are given by the column vector

$$S_0 = \left(S_0^1, S_0^2, \ldots, S_0^N \right)^t \triangleq \begin{pmatrix} S_0^1 \\ S_0^2 \\ \vdots \\ S_0^N \end{pmatrix}.$$

Notation   For vectors and matrices we shall use the superscript 't' to denote transpose.

Uncertainty about the market is represented by a finite number of possible states in which the market might be at time T that we label $1, 2, \ldots, n$. The security values at time T are given by an $N \times n$ matrix $D = (D_{ij})$, where the coefficient D_{ij} is the value of the ith security at time T if the market is in state j. Our binary model corresponds to $N = 2$ (the stock and a riskless cash bond) and $n = 2$ (the two states being determined by the two possible values of S_T).

In this notation, a portfolio can be thought of as a vector $\theta = (\theta_1, \theta_2, \ldots, \theta_n)^t \in \mathbb{R}^N$, whose market value at time zero is the scalar product $S_0 \cdot \theta = S_0^1 \theta_1 + S_0^2 \theta_2 + \cdots + S_0^N \theta_N$. The value of the portfolio at time T is a vector in $\mathbb{R}^n$ whose ith entry is the value of the portfolio if the market is in state i. We can write the value at time T as

$$\begin{pmatrix} D_{11}\theta_1 + D_{21}\theta_2 + \cdots + D_{N1}\theta_N \\ D_{12}\theta_1 + D_{22}\theta_2 + \cdots + D_{N2}\theta_N \\ \vdots \\ D_{1n}\theta_1 + D_{2n}\theta_2 + \cdots + D_{Nn}\theta_N \end{pmatrix} = D^t \theta.$$

> **Notation**   For a vector $x \in \mathbb{R}^n$ we write $x \geq 0$, or $x \in \mathbb{R}^n_+$, if $x = (x_1, \ldots, x_n)$ and $x_i \geq 0$ for all $i = 1, \ldots, n$. We write $x > 0$ to mean $x \geq 0$, $x \neq 0$. Notice that $x > 0$ does not require x to be strictly positive in *all* its coordinates. We write $x \gg 0$, or $x \in \mathbb{R}^n_{++}$, for vectors in $\mathbb{R}^n$ that are *strictly* positive in *all* coordinates.

In this notation, an *arbitrage* is a portfolio $\theta \in \mathbb{R}^N$ with either

$$S_0 \cdot \theta \leq 0, \quad D^t \theta > 0 \qquad \text{or} \qquad S_0 \cdot \theta < 0, \quad D^t \theta \geq 0.$$

Arbitrage pricing

The key to arbitrage pricing in this model is the notion of a state price vector.

Definition 1.5.1   *A* state price vector *is a vector $\psi \in \mathbb{R}^n_{++}$ such that $S_0 = D\psi$.*

To see why this terminology is natural, we first expand this to obtain

$$\begin{pmatrix} S_0^1 \\ S_0^2 \\ \vdots \\ S_0^N \end{pmatrix} = \psi_1 \begin{pmatrix} D_{11} \\ D_{21} \\ \vdots \\ D_{N1} \end{pmatrix} + \psi_2 \begin{pmatrix} D_{12} \\ D_{22} \\ \vdots \\ D_{N2} \end{pmatrix} + \cdots + \psi_n \begin{pmatrix} D_{1n} \\ D_{2n} \\ \vdots \\ D_{Nn} \end{pmatrix}. \qquad (1.2)$$

The vector, $D^{(i)}$, multiplying ψ_i is the security price vector if the market is in state i. We can think of ψ_i as the marginal cost at time zero of obtaining an additional unit of wealth at the end of the time period if the system is in state i. In other words, if at the end of the time period, the market is in state i, then the value of our portfolio increases by one for each additional ψ_i of investment at time zero. To see this, suppose that we can find vectors $\{\theta^{(i)} \in \mathbb{R}^N\}_{1 \leq i \leq n}$ such that

$$\theta^{(i)} \cdot D^{(j)} = \begin{cases} 1 & \text{if } i = j, \\ 0 & \text{otherwise.} \end{cases}$$

That is, the value of the portfolio $\theta^{(i)}$ at time T is the indicator function that the market is in state i. Then, using equation (1.2), the cost of purchasing $\theta^{(i)}$ at time zero is precisely $S_0 \cdot \theta^{(i)} = \left(\sum_{j=1}^n \psi_j D^{(j)} \right) \cdot \theta^{(i)} = \psi_i$. Such portfolios $\{\theta^{(i)}\}_{1 \leq i \leq n}$ are called *Arrow–Debreu securities*.

We shall find a convenient way to think about the state price vector in §1.6, but first, here is the key result.

Theorem 1.5.2   *For the market model described above there is no arbitrage if and only if there is a state price vector.*

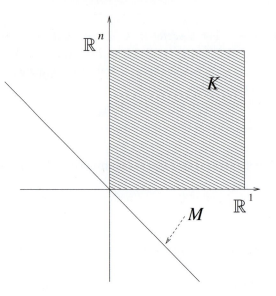

Figure 1.4   There is no arbitrage if and only if the regions K and M of Theorem 1.5.2 intersect only at the origin.

This result, due to Harrison & Kreps (1979), is the simplest form of what is often known as the Fundamental Theorem of Asset Pricing. The proof is an application of a Hahn–Banach Separation Theorem, sometimes called the *Separating Hyperplane Theorem*. We shall also need the *Riesz Representation Theorem*. Recall that $M \subseteq \mathbb{R}^d$ is a *cone* if $x \in M$ implies $\lambda x \in M$ for all strictly positive scalars λ and that a *linear functional* on $\mathbb{R}^d$ is a linear mapping $F\colon \mathbb{R}^d \to \mathbb{R}$.

Theorem 1.5.3 (Separating Hyperplane Theorem)   *Suppose M and K are closed convex cones in $\mathbb{R}^d$ that intersect precisely at the origin. If K is* not *a linear subspace, then there is a non-zero linear functional F such that $F(x) < F(y)$ for each $x \in M$ and each non-zero $y \in K$.*

This version of the Separating Hyperplane Theorem can be found in Duffie (1992).

Theorem 1.5.4 (Riesz Representation Theorem)   *Any bounded linear functional on $\mathbb{R}^d$ can be written as $F(x) = v_0 \cdot x$. That is $F(x)$ is the scalar product of some fixed vector $v_0 \in \mathbb{R}^d$ with x.*

Proof of Theorem 1.5.2: We take $d = 1 + n$ in Theorem 1.5.3 and set

$$M = \left\{ \left(-S_0 \cdot \theta, D^t \theta \right) : \theta \in \mathbb{R}^N \right\} \subseteq \mathbb{R} \times \mathbb{R}^n = \mathbb{R}^{1+n},$$

$$K = \mathbb{R}_+ \times \mathbb{R}_+^n.$$

Note that K is a cone and not a linear space, M is a linear space. Evidently, there is no arbitrage if and only if K and M intersect precisely at the origin as shown in

Figure 1.4. We must prove that $K \cap M = \{0\}$ if and only if there is a state price vector.

(i) *Suppose first that $K \cap M = \{0\}$.* From Theorem 1.5.3, there is a linear functional $F \colon \mathbb{R}^d \to \mathbb{R}$ such that $F(z) < F(x)$ for all $z \in M$ and non-zero $x \in K$.

The first step is to show that F must vanish on M. We exploit the fact that M is a linear space. First observe that $F(0) = 0$ (by linearity of F) and $0 \in M$, so $F(x) \geq 0$ for $x \in K$ and $F(x) > 0$ for $x \in K \backslash \{0\}$. Fix $x_0 \in K$ with $x_0 \neq 0$. Now take an arbitrary $z \in M$. Then $F(z) < F(x_0)$, but also, since M is a linear space, $\lambda F(z) = F(\lambda z) < F(x_0)$ for all $\lambda \in \mathbb{R}$. This can only hold if $F(z) = 0$. $z \in M$ was arbitrary and so F vanishes on M as required.

We now use this actually to construct explicitly the state price vector from F. First we use the Riesz Representation Theorem to write F as $F(x) = v_0 \cdot x$ for some $v_0 \in \mathbb{R}^d$. It is convenient to write $v_0 = (\alpha, \phi)$ where $\alpha \in \mathbb{R}$ and $\phi \in \mathbb{R}^n$. Then

$$F(v, c) = \alpha v + \phi \cdot c \qquad \text{for any } (v, c) \in \mathbb{R} \times \mathbb{R}^n = \mathbb{R}^d.$$

Since $F(x) > 0$ for all non-zero $x \in K$, we must have $\alpha > 0$ and $\phi \gg 0$ (consider a vector along each of the coordinate axes). Finally, since F vanishes on M,

$$-\alpha S_0 \cdot \theta + \phi \cdot D^t \theta = 0 \qquad \text{for all } \theta \in \mathbb{R}^N.$$

Observing that $\phi \cdot D^t \theta = (D\phi) \cdot \theta$, this becomes

$$-\alpha S_0 \cdot \theta + (D\phi) \cdot \theta = 0 \qquad \text{for all } \theta \in \mathbb{R}^N,$$

which implies that $-\alpha S_0 + D\phi = 0$. In other words, $S_0 = D(\phi/\alpha)$. The vector $\psi = \phi/\alpha$ is a state price vector.

(ii) *Suppose now that there is a state price vector, ψ.* We must prove that $K \cap M = \{0\}$. By definition, $S_0 = D\psi$ and so for any portfolio θ,

$$S_0 \cdot \theta = (D\psi) \cdot \theta = \psi \cdot (D^t \theta). \tag{1.3}$$

Suppose that for some portfolio θ, $(-S_0 \cdot \theta, D^t \theta) \in K$. Then $D^t \theta \in \mathbb{R}^n_+$ and $-S_0 \cdot \theta \geq 0$. But since $\psi \gg 0$, if $D^t \theta \in \mathbb{R}^n_+$, then $\psi \cdot (D^t \theta) \geq 0$ which, by equation (1.3), tells us that $S_0 \cdot \theta \geq 0$. Thus it must be that $S_0 \cdot \theta = 0$ *and* $D^t \theta = 0$. That is, $K \cap M = \{0\}$, as required.   □

1.6 The risk-neutral probability measure

The state price vector then is the key to arbitrage pricing for our multiasset market models. Although we have an economic interpretation for it, in order to pave the way for the full machinery of probability and martingales we must think about it in a different way.

Recall that all the entries of ψ are strictly positive.

State prices and probability

Writing $\psi_0 = \sum_{i=1}^n \psi_i$, we can think of

$$\underline{\psi} \triangleq \left(\frac{\psi_1}{\psi_0}, \frac{\psi_2}{\psi_0}, \ldots, \frac{\psi_n}{\psi_0}\right)^t \tag{1.4}$$

as a vector of *probabilities* for being in different states. It is important to emphasise that they may have nothing to do with our view of how the markets will move. First of all,

What is ψ_0?

Suppose that as in our binary model (where we had a risk-free cash bond) the market allows *positive riskless borrowing*. In this general setting we just suppose that we can replicate such a bond by a portfolio $\overline{\theta}$ for which

$$D^t\overline{\theta} = \begin{pmatrix} 1 \\ 1 \\ \vdots \\ 1 \end{pmatrix},$$

i.e. the value of the portfolio at time T is one, no matter what state the market is in. Using the fact that ψ is a state price vector, we calculate that the cost of such a portfolio at time zero is

$$S_0 \cdot \overline{\theta} = (D\psi) \cdot \overline{\theta} = \psi \cdot (D^t\overline{\theta}) = \sum_{i=1}^n \psi_i = \psi_0.$$

That is ψ_0 *represents the discount on riskless borrowing.* In our notation of §1.2, $\psi_0 = e^{-rT}$.

Expectation recovered

Now under the probability distribution given by the vector (1.4), the expected value of the ith security at time T is

$$\mathbb{E}[S_T^i] = \sum_{j=1}^n D_{ij} \frac{\psi_j}{\psi_0} = \frac{1}{\psi_0} \sum_{j=1}^n D_{ij}\psi_j = \frac{1}{\psi_0} S_0^i,$$

where in the last equality we have used $S_0 = D\psi$. That is

$$S_0^i = \psi_0 \mathbb{E}[S_T^i], \qquad i = 1, \ldots, n. \tag{1.5}$$

Any security's price is its discounted expected payoff under the probability distribution (1.4). The same must be true of any portfolio. This observation gives us a new way to think about the pricing of contingent claims.

Definition 1.6.1   *We shall say that a claim, C, at time T is attainable if it can be hedged. That is, if there is a portfolio whose value at time T is exactly C.*

Notation When we wish to emphasise the underlying probability measure, $\mathbb{Q}$, we write $\mathbb{E}^{\mathbb{Q}}$ for the expectation operator.

Theorem 1.6.2   *If there is no arbitrage, the unique time zero price of an attainable claim C at time T is $\psi_0 \mathbb{E}^{\mathbb{Q}}[C]$ where the expectation is with respect to any probability measure $\mathbb{Q}$ for which $S_0^i = \psi_0 \mathbb{E}^{\mathbb{Q}}[S_T^i]$ for all i and ψ_0 is the discount on riskless borrowing.*

Remark: Notice that it is crucial that the claim is attainable (see Exercise 11).   □

Proof of Theorem 1.6.2: By Theorem 1.5.2 there is a state price vector and this leads to the probability measure (1.4) satisfying $S_0^i = \psi_0 \mathbb{E}\left[S_T^i\right]$ for all i. Since the claim can be hedged, there is a portfolio θ such that $\theta \cdot S_T = C$. In the absence of arbitrage, the time zero price of the claim is the cost of this portfolio at time zero,

$$\theta \cdot S_0 = \theta \cdot (\psi_0 \mathbb{E}[S_T]) = \psi_0 \sum_{i=1}^{N} \theta_i \mathbb{E}[S_T^i] = \psi_0 \mathbb{E}[\theta \cdot S_T].$$

The same value is obtained if the expectation is calculated for any vector of probabilities, $\mathbb{Q}$, such that $S_0^i = \psi_0 \mathbb{E}^{\mathbb{Q}}\left[S_T^i\right]$ since, in the absence of arbitrage, there is only one riskless borrowing rate and this completes the proof.   □

Risk-neutral pricing   In this language, our arbitrage pricing result says that if we can find a probability vector for which the time zero value of each underlying security is its discounted expected value at time T then we can find the time zero value of any *attainable* contingent claim by calculating its discounted expectation. Notice that we use the *same* probability vector, whatever the claim.

Definition 1.6.3   *If our market can be in one of n possible states at time T, then any vector, $p = (p_1, p_2, \ldots, p_n) \gg 0$, of probabilities for which each security's price is its discounted expected payoff is called a* risk-neutral probability measure *or* equivalent martingale measure.

The term *equivalent* reflects the condition that $p \gg 0$; cf. Definition 2.3.12. Our simple form of the Fundamental Theorem of Asset Pricing (Theorem 1.5.2) says that in a market with positive riskless borrowing there is no arbitrage if and only if there is an equivalent martingale measure. We shall refer to the process of pricing by taking expectations with respect to a risk-neutral probability measure as *risk-neutral pricing*.

Example 1.3.1 revisited Let us return to our very first example of pricing a European call option and confirm that the above formula really does give us the arbitrage price.
    Here we have just two securities, a cash bond and the underlying stock. The discount on borrowing is $\psi_0 = e^{-rT}$, but we are assuming that the Yen interest rate is zero, so $\psi_0 = 1$. The matrix of security values at time T is given by

$$D = \begin{pmatrix} 1 & 1 \\ 4000 & 2000 \end{pmatrix}.$$

Writing p for the risk-neutral probability that the security price vector is $(1, 4000)'$, if the stock price is to be equal to its discounted expected payoff, p must solve

$$4000p + 2000(1 - p) = 2500,$$

which gives $p = 0.25$. The contingent claim is ¥1000 if the stock price at expiry is ¥4000 and zero otherwise. The expected value of the claim under the risk-neutral probability, and therefore (since interest rates are zero) the price of the option, is then ¥0.25 × 1000 = ¥250, as before.

An advantage of this approach is that, armed with the probability p, it is now a trivial matter to price all European options on this stock with the same expiry date (six months time) by taking expectations with respect to the *same* probability measure. For example, for a European put option with strike price ¥3500, the price is

$$¥\mathbb{E}\left[(K - S_T)_+\right] = ¥0.75 \times 1500 = ¥1125.$$

Our original argument would lead to a new set of simultaneous equations for each new claim.   □

Complete markets

We now have a prescription for the arbitrage price of a claim if one exists, that is if the claim is attainable. But we must be a little cautious. Arbitrage prices only exist for attainable claims – even though the prescription may continue to make sense.

Definition 1.6.4   *A market is said to be* complete *if every contingent claim is attainable, i.e. if every possible derivative claim can be hedged.*

Proposition 1.6.5   *A market consisting of N tradable assets, evolving according to a single period model in which at the end of the time period the market is one of n possible states, is complete if and only if $N \geq n$ and the rank of the matrix, D, of security prices is n.*

Proof: Any claim in our market can be expressed as a vector $v \in \mathbb{R}^n$. A hedge for that claim will be a portfolio $\theta = \theta(v) \in \mathbb{R}^N$ for which $D'\theta = v$. Finding such a θ amounts to solving n equations in N unknowns. Thus a hedging portfolio exists for *every* choice of v if and only if $N \geq n$ and the rank of D is n, as required.   □

Notice in particular that our single period binary model is complete.

Suppose that our market is complete and arbitrage-free and let $\mathbb{Q}$ and $\mathbb{Q}'$ be any two equivalent martingale measures. By completeness every claim is attainable, so for *every* random variable X, using that there is only one risk-free rate,

$$\mathbb{E}^{\mathbb{Q}}[X] = \mathbb{E}^{\mathbb{Q}'}[X].$$

In other words $\mathbb{Q} = \mathbb{Q}'$. So in a complete arbitrage-free market the equivalent martingale measure is *unique*.

The main
results so far

Let us summarise the results for our single period markets. They will be reflected again and again in what follows.

Results for single period models
- The market is arbitrage-free if and only if there exists a martingale measure, $\mathbb{Q}$.
- The market is complete if and only if $\mathbb{Q}$ is unique.
- The arbitrage price of an attainable claim C is $e^{-rT}\mathbb{E}^{\mathbb{Q}}[C]$.

Martingale measures are a powerful tool. However, in an incomplete market, if a claim C is not attainable different martingale measures can give different prices. The arbitrage-free notion of *fair price* only makes sense if we can *hedge*.

Trading in
two different
markets

We must sound just one more note of caution. It is important in calculating the risk-neutral probabilities that all the assets being modelled are tradable in the same market. We illustrate with an example.

Example 1.6.6   *Suppose that in the US dollar markets the current Sterling exchange rate is* 1.5 *(so that £100 costs $150). Consider a European call option that offers the holder the right to buy* £100 *for $150 at time* T. *The riskless borrowing rate in the UK is* u *and that in the US is* r. *Assuming a single period binary model in which the exchange rate at the expiry time is either* 1.65 *or* 1.45, *find the fair price of this option.*

Solution: Now we have a problem. The exchange rate is *not* tradable. Nor, in *dollar* markets, is a Sterling cash bond – it is a tradable instrument, but in Sterling markets. However, the product of the two *is* a dollar tradable and we shall denote the value of this product by S_t at time t.

Now, since the riskless interest rate in the UK is u, the time zero price of a Sterling cash bond, promising to pay £1 at time T, is e^{-uT} and, of course, at time T the bond price is one. Thus we have $S_0 = e^{-uT}150$ and $S_T = 165$ or $S_T = 145$.

Let p be the risk-neutral probability that $S_T = 165$. Then, since the discounted price (in the *dollar* market) of our 'asset' at time T must have expectation S_0, we obtain

$$150e^{-uT} = e^{-rT}(165p + 145(1-p)),$$

which yields

$$p = \frac{150e^{(r-u)T} - 145}{20}.$$

The price of the option is the discounted expected payoff with respect to this probability which gives

$$V_0 = e^{-rT} 15p = \frac{3}{4}\left(150e^{-uT} - 145e^{-rT}\right).$$

□

Exercises

1 What view about the market is reflected in each of the following strategies?

   (a) *Bullish vertical spread:* Buy one European call and sell a second one with the same expiry date, but a larger strike price.
   (b) *Bearish vertical spread:* Buy one European call and sell a second one with the same expiry date but a smaller strike price.
   (c) *Strip:* Buy one European call and two European puts with the same exercise date and strike price.
   (d) *Strap:* Buy two European calls and one European put with the same exercise date and strike price.
   (e) *Strangle:* Buy a European call and a European put with the same expiry date but different strike prices (consider all possible cases).

2 A *butterfly spread* represents the complementary bet to the straddle. It has the following payoff at expiry:

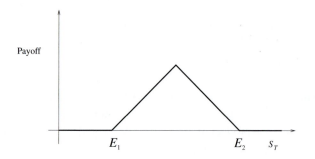

Find a portfolio consisting of European calls and puts, all with the same expiry date, that has this payoff.

3 Suppose that the price of a certain asset has the lognormal distribution. That is $\log(S_T/S_0)$ is normally distributed with mean v and variance σ^2. Calculate $\mathbb{E}[S_T]$.

4 (a) Prove Lemma 1.3.2.
   (b) What happens if we drop the assumption that $d < e^{rT} < u$?

5 Suppose that at current exchange rates, £100 is worth €160. A speculator believes that by the end of the year there is a probability of $1/2$ that the pound will have fallen to €1.40, and a $1/2$ chance that it will have gained to be worth €2.00. He therefore buys a European put option that will give him the right (but not the obligation) to

sell £100 for €1.80 at the end of the year. He pays €20 for this option. Assume that the risk-free interest rate is zero across the Euro-zone. Using a single period binary model, either construct a strategy whereby one party is certain to make a profit or prove that this is the fair price.

6   How should we modify the analysis of Example 1.3.1 if we are pricing an option based on a commodity such as oil?

7   Show that if there is no arbitrage in the market, then any portfolio constructed at time zero that exactly replicates a claim C at time T has the same value at time zero.

8   *Put–call parity:* Denote by C_t and P_t respectively the prices at time t of a European call and a European put option, each with maturity T and strike K. Assume that the risk-free rate of interest is constant, r, and that there is no arbitrage in the market. Show that for each $t \le T$,

$$C_t - P_t = S_t - Ke^{-r(T-t)}.$$

9   Use risk-neutral pricing to value the option in Exercise 5. Check your answer by constructing a portfolio that exactly replicates the claim at the expiry of the contract.

10   What is the payoff of a forward at expiry? Use risk-neutral pricing to solve the pricing problem for a forward contract.

11   Consider the *ternary* model for the underlying of §1.4. How many equivalent martingale measures are there? If there are two different martingale measures, do they give the same price for a claim? Are there arbitrage opportunities?

12   Suppose that the value of a certain stock at time T is a random variable with distribution $\mathbb{P}$. Note we are *not* assuming a binary model. An option written on this stock has payoff C at time T. Consider a portfolio consisting of ϕ units of the underlying and ψ units of bond, held until time T, and write V_0 for its value at time zero. Assuming that interest rates are zero, show that the extra cash required by the holder of this portfolio to meet the claim C at time T is

$$\Psi \triangleq C - V_0 - \phi\,(S_T - S_0)\,.$$

Find expressions for the values of V_0 and ϕ (in terms of $\mathbb{E}[S_T]$, $\mathbb{E}[C]$, $var[S_T]$ and $cov\,(S_T, C)$) that minimise

$$\mathbb{E}[\Psi^2],$$

and check that for these values $\mathbb{E}[\Psi] = 0$.
Prove that for a binary model, any claim C depends *linearly* on $S_T - S_0$. Deduce that in this case we can find V_0 and ϕ such that $\Psi = 0$.
When the model is *not* complete, the parameters that minimise $\mathbb{E}\left[\Psi^2\right]$ correspond to finding the best linear approximation to C (based on $S_T - S_0$). The corresponding value of the expectation is a measure of the *intrinsic risk* in the option.

13 *Exchange rate forward:* Suppose that the riskless borrowing rate in the UK is u and that in the USA is r. A dollar investor wishes to set the exchange rate, C_T, in a forward contract in which the two parties agree to exchange C_T dollars for one pound at time T. If a pound is currently C_0 dollars, what is the fair value of C_T?

14 The option writer in Example 1.6.6 sells a *digital* option to a speculator. This amounts to a bet that the asset price will go up. The payoff is a fixed amount of cash if the exchange rate goes to \$165 per £100, and nothing if it goes down. If the speculator pays \$10 for this bet, what cash payout should the option writer be willing to write into the option? You may assume that interest rates are zero.

15 Suppose now that the seller of the option in Example 1.6.6 operates in the Sterling markets. Reexpress the market in terms of Sterling tradables and find the corresponding risk-neutral probabilities. Are they the same as the risk-neutral probabilities calculated by the dollar trader? What is the dollar cost at time zero of the option as valued by the Sterling trader?
    This is an example of *change of numeraire*. The dollar trader uses the dollar bond as the reference risk-free asset whereas the Sterling trader uses a Sterling bond.

2 Binomial trees and discrete parameter martingales

Summary

In this chapter we build some more sophisticated market models that track the evolution of stock prices over a succession of time periods. Over each individual time period, the market follows our simple binary model of Chapter 1. The possible trajectories of the stock prices are then encoded in a tree. A simple corollary of our work of Chapter 1 will allow us to price claims by taking expectation with respect to certain probabilities on the tree under which the stock price process is a discrete parameter *martingale*.

Definitions and basic properties of discrete parameter martingales are presented and illustrated in §2.3, and we see for the first time how martingale methods can be employed as an elegant computational tool. Then, §2.4 presents some important martingale theorems. In §2.5 we pave the way for the Black–Scholes analysis of Chapter 5 by showing how to construct, in the martingale framework, the portfolio that replicates a claim. In §2.6 we preview the Black–Scholes formula with a heuristic passage to the limit.

2.1 The multiperiod binary model

Our single period binary model is, of course, inadequate as a model of the evolution of an asset price. In particular, we have allowed ourselves to observe the market at just two times, zero and T. Moreover, at time T, we have supposed the stock price to take one of just two possible values. In this section we construct more sophisticated market models by stringing together copies of our single period model into a tree.

Once again our financial market will consist of just two instruments, the stock and a cash bond. As before we assume that unlimited amounts of both can be bought and sold without transaction costs. There is no risk of default on a promise and the market is prepared to buy and sell a security for the same price (that is, there is no *bid–offer spread*).

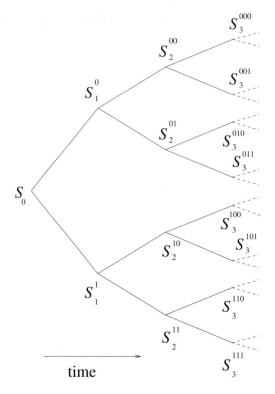

S_3^{000}

S_2^{00}

S_3^{001}

S_1^0

S_2^{01}

S_3^{010}

S_3^{011}

S_0

S_3^{100}

S_3^{101}

S_2^{10}

S_1^1

S_3^{110}

S_2^{11}

time

S_3^{111}

Figure 2.1   The tree of stock prices.

We suppose the market to be observable at times $0 = t_0 < t_1 < \cdots < t_N = T$.

The stock

Over each time period $[t_i, t_{i+1}]$ the stock follows the binary model. This is illustrated in Figure 2.1. After i time periods, the stock can have any of 2^i possible values. However, *given* its value at time t_i there are only *two* admissible possibilities for the stock price at time t_{i+1}. It is not necessary, but it is conventional, to suppose that all time periods have the same length and so we shall write $t_i = i\delta t$ where $\delta t = T/N$.

The cash bond

In our simple model, the cash bond behaved entirely predictably. There was a known interest rate, r, and the cash bond increased in value over a time period of length T by a factor e^{rT}. Now, we do not have to impose such a stringent condition. The interest rate can itself be random, varying over different time periods. Our work will generalise immediately provided that we insist that the interest rate over the time interval $[t_i, t_{i+1})$ is known at the *start* of that interval, although it may depend on which of the 2^i nodes our market is in. In this way, we admit the possibility of randomness in our cash bond. Notice however that it is a very different sort of randomness from that of the stock. The value of the bond at time t_{i+1} is already known to us at time t_i. This is certainly not true for the stock. In spite of our new-found freedom, for simplicity, we shall continue to suppose that the interest rate is the constant, r.

Replicating portfolios

At first sight it is not clear that we can make progress with our new model. For a tree consisting of k time steps there are 2^k possible values for the stock price. If we now look back at Proposition 1.6.5, this suggests that we need at least 2^k stocks to be traded in our market if we want it to be complete. For $k = 20$, this requires over a million 'independent' assets, far more than we see in any real market. But things are not so bad. More claims become attainable if we allow ourselves to rebalance our replicating portfolio after each time period. The only restriction that we impose is that this rebalancing cannot involve any extra input of cash: the purchase of more stock must be funded by the sale of some of our bonds and vice versa. This will be formalised later as the *self-financing* property.

Backwards induction on the tree

The key to understanding pricing and hedging in this bigger model is *backwards induction* on the tree of stock prices.

Example 2.1.1 (Pricing a European call)   *Suppose again that we are pricing a European option with maturity time T. As above, we set $\delta t = T/N$ so that T corresponds to N time periods and we write S_i for the stock price at time $i\delta t$. The payoff of the option at time T is denoted by C_N.*

Method: The key idea is as follows. Suppose that we *know* the price, S_{N-1}, of the stock at time $(N-1)\delta t$. Then our previous analysis would tell us the value, C_{N-1}, of the option at time $(N-1)\delta t$. Specifically, $C_{N-1} = \psi_0^{(N)}\mathbb{E}_{N-1}[C_N]$ where the expectation is with respect to a probability measure for which $S_{N-1} = \psi_0^{(N)}\mathbb{E}_{N-1}[S_N]$ and $\psi_0^{(N)} = e^{-r\delta t}$. (In a world of varying interest rates r must be replaced by the rate at the node of the tree corresponding to the *known* value of S_{N-1}.) Moreover, using Lemma 1.3.2, we know how to construct a portfolio at time $(N-1)\delta t$ that will have value exactly C_N at time $N\delta t$. In this way, for each of the 2^{N-1} nodes of the tree at time $(N-1)\delta t$, we calculate the amount of money, C_{N-1}, that we require to construct a portfolio that exactly replicates the claim C_N at time T.

We now think of C_{N-1} as a *claim* at time $(N-1)\delta t$ and we repeat the process. If we *know* S_{N-2}, we can construct a portfolio at time $(N-2)\delta t$ whose value at time $(N-1)\delta t$ will be exactly C_{N-1}, and this portfolio will cost us $\psi_0^{(N-1)}\mathbb{E}_{N-2}[C_{N-1}]$, where the expectation is with respect to a measure such that $S_{N-2} = \psi_0^{(N-1)}\mathbb{E}_{N-2}[S_{N-1}]$. Here again $\psi_0^{(N-1)} = e^{-r\delta t}$. Continuing in this way, we successively calculate the cost of a portfolio that, after appropriate readjustment at each tick of the clock, but without any extra input of wealth and without paying dividends, will allow us to meet exactly the claim against us at time $N\delta t = T$. We'll illustrate the method in Example 2.1.2. ☐

Binomial trees

It is useful to consider a special form of the binary tree in which over each time step $[t_i, t_{i+1}]$ the stock price either increases from its current value, S_i, to S_iu or decreases to S_id for some constants $0 < d < u < \infty$. In such a tree the same stock price can be attained in many different ways. For example the value S_0ud at time t_2 can be attained as the result of an upward stock movement followed by a downward stock

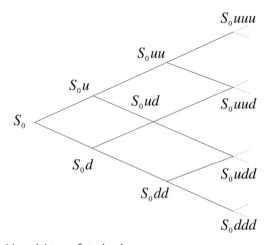

Figure 2.2   A recombinant or binomial tree of stock prices.

movement or vice versa. The tree of stock prices then takes the form of Figure 2.2. Such a tree is said to be *recombinant* (different branches can recombine). These special recombinant trees are also known as *binomial trees* since (provided u, d, and r remain constant over time) the risk-neutral probability measure will be the same on each upward branch and so the stock price at time $t_n = n\delta t$ is determined by a binomial distribution. Such trees are computationally much easier to work with than general binary trees and, as we shall see, are quite adequate for our purposes. The binomial model was introduced by Cox, Ross & Rubinstein (1979) and has played a key rôle in the derivatives industry.

   We now illustrate the method of backwards induction on a recombinant tree.

Example 2.1.2   *Suppose that stock prices are given by the tree in Figure 2.3 and that $\delta t = 1$. If interest rates are zero, what is cost of an option to buy the stock at price* 100 *at time* 3?

Solution: It is easy to fill in the value of the claim at time 3. Reading from top to bottom, the claim has values 60, 20, 0 and 0.

   Next we need to find the risk-neutral probabilities for each triad of nodes of the form

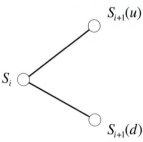

Evidently in this example the risk-neutral probability of stepping up is $1/2$ at every node. We can now calculate the value of the option at the penultimate time, 2, to be,

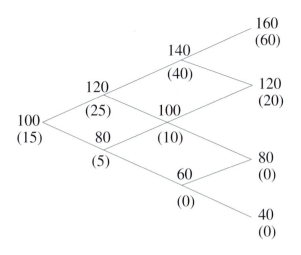

Figure 2.3   The tree of stock prices for the underlying stock in Example 2.1.2. The number in brackets is
the value of the claim at each node.

again reading downward, 40, 10, 0. Repeating this for time 1 gives values 25 (if the
price steps up from time 0) and 5 (if the price has stepped down). Finally, then, the
value of the option at time 0 is 15.

Having filled in the option prices on the tree, we can now construct a portfolio
that exactly replicates the claim at time 3 using the prescription of Lemma 1.3.2. We
write (ϕ_i, ψ_i) for the amount of stock and bond held in the portfolio over the time
interval $[(i-1)\delta t, i\delta t)$.

- At time 0, we are given 15 for the option. We calculate ϕ_1 as $(25-5)/(120-80) =$
  0.5. So we buy 0.5 units of stock, which costs 50, and we borrow 35 in cash bonds.
- Suppose that $S_1 = 120$. The new ϕ is $(40-10)/(140-100) = 0.75$, so we buy
  another 0.25 units of stock, taking our total bond borrowing to 65.
- Suppose that $S_2 = 140$. Now $\phi = (60-20)/(160-120) = 1$, so we buy still more
  stock, to take our holding up to 1 unit and our total borrowing to 100 bonds.
- Finally, suppose that $S_3 = 120$. The option will be in the money, so we must hand
  over our unit of stock for 100, which is exactly enough to cancel our bond debt.

The table below summarises our stock and bond holding if the stock price follows
another path through the tree.

Time i	Last jump	Stock price S_i	Option value V_i	Stock holding ϕ_i	Bond holding ψ_i
0	—	100	15	—	—
1	down	80	5	0.50	−35
2	up	100	10	0.25	−15
3	down	80	0	0.50	−40

Notice that all of the processes $\{S_i\}_{0\leq i\leq N}$, $\{V_i\}_{0\leq i\leq N}$, $\{\phi_i\}_{1\leq i\leq N}$, $\{\psi_i\}_{1\leq i\leq N}$ depend on the sequence of up and down jumps. In particular, $\{\phi_i\}_{1\leq i\leq N}$ and $\{\psi_i\}_{1\leq i\leq N}$ are random too. We do *not* know the dynamics of the portfolio at time 0. However, we *do* know that our portfolio is *self-financing*. The portfolio that we hold over $[i+1, i+2)$ can be bought with the proceeds of liquidating (at time $i + 1$) the portfolio that we held over the time interval $[i, i + 1)$ – there is no need for any extra input of cash. Moreover, we know how to adjust our portfolio at each time step on the basis of knowledge of the *current* stock price. There is no risk.   □

In the single period binary model, we saw that any claim at time T was attainable and its price at time zero could be expressed as an expectation. The same is true in the multiperiod setting (see Exercise 1). The proof that any claim is attainable is just backwards induction on the tree. To recover the pricing formula as an expectation, we define a probability distribution on paths through the tree.

Path probabilities   Notice that our backwards induction argument has specified exactly one probability on each branch of the tree. For each *path* through the tree that the stock price could follow we define the path probability to be the product of the probabilities on the branches that comprise it.

In Exercise 2 you are asked to show that the price of a claim at time T that we obtained by backwards induction is precisely the discounted expected value of the claim with respect to these path probabilities (in which the discounted claim at each node is weighted according to the sum of the probabilities of all paths that end at that node). Let's just check this prescription for our preceding example. In the recombining tree of Example 2.1.2, there are a total of eight paths, one ending at the top node, one at the bottom and three at each of the other nodes. Each path has equal probability, $1/8$, and the expectation of the claim is therefore $1/8 \times 60 + 3/8 \times 20 = 15$, which is the price that we calculated by backwards induction.

2.2   American options

Our somewhat more sophisticated market model is sufficient for us to take a first look at options whose payoff depends on the *path* followed by the stock price over the time interval $[0, T]$. In this section we concentrate on the most important examples of such options: American options.

Definition 2.2.1 (American calls and puts)   *An* American call option *with strike price K and expiry time T gives the holder the right, but not the obligation, to* buy *an asset for price K at* any time up to T.

An American put option *with strike price K and expiry time T gives the holder the right, but not the obligation, to* sell *an asset for price K at* any time up to T.

Evidently the value of an American option should be more than (or at least no less than) that of its European counterpart. The question is, how much more?

Calls on non-dividend-paying stock

First let us prove the following oft-quoted result.

Lemma 2.2.2 *It is never optimal to exercise an American call option on non-dividend-paying stock before expiry.*

Proof: Consider the following two portfolios.

- **Portfolio A:** One American call option plus an amount of cash equal to $Ke^{-r(T-t)}$ at time t.
- **Portfolio B:** One share.

Writing S_t for the share price at time t, if the call option is exercised at time $t < T$, then the value of portfolio **A** at time t is $S_t - K + Ke^{-r(T-t)} < S_t$. (Evidently the option will only be exercised if $S_t > K$.) The value of portfolio **B** is S_t. On the other hand at time T, if the option is exercised then the value of portfolio **A** is $\max\{S_T, K\}$ which is *at least* that of portfolio **B**.

We have shown that exercising prior to maturity gives a portfolio whose value is less than that of portfolio **B** whereas exercising at maturity gives a portfolio whose value is greater than or equal to that of **B**. It cannot be optimal to exercise early. □

This result only holds for non-dividend-paying stock. An alternative proof of Lemma 2.2.2 is Exercise 5. In Exercise 7 the result is extended to show that if the underlying stock pays discrete dividends, then it can only be optimal to exercise at the final time T or at one of the dividend times (see also Exercise 8). More generally, the decision whether to exercise early depends on the 'cost' in terms of lost dividend income.

Put on non-dividend-paying stock

The case of American put options is harder (even without dividends). We illustrate with an example.

Example 2.2.3 *Suppose once again that our asset price evolves according to the recombinant tree of Figure 2.3. To illustrate the method, again we suppose that the risk-free interest rate is zero (but see the second paragraph of Remark 2.2.4). What is the value of a three month American put option with strike price 100?*

Solution: As in the case of a European option, we work our way backwards through the tree.

- The value of the claim at time 3, reading from top to bottom, is 0, 0, 20, 60.
- At time 2, we must consider two possibilities: the value if we exercise the claim, and the value if we do not. For the top node it is easy. The value is zero either way. For the second node, the stock price is equal to the strike price, so the value is zero if we exercise the option. On the other hand, if we don't, then from our analysis of the single step binary model, the value of the claim is the expected value under the risk-neutral probabilities of the claim at time 3. We already calculated the risk-neutral

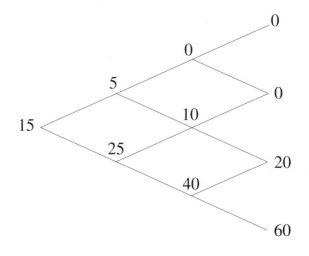

Figure 2.4   The evolution of the price of the American put option of Example 2.2.3.

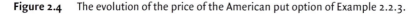

probabilities to be $1/2$ on each branch of the tree, so this expected value is 10. For the bottom node, the value is 40 whether or not we exercise the claim.

- Now consider the two nodes at time 1. For the top one, if we exercise the option it is worthless whereas if we hold it then, again by our analysis of the single period model, its value is 5. For the bottom node, if we exercise the option then it is worth 20, whereas if we wait it is worth 25.
- Finally, at time 0, if we exercise, the value is zero, whereas if we wait the value is 15.

The option prices are shown in Figure 2.4.

□

Remark 2.2.4

1   Notice that in the above example it was not optimal to exercise the option at time 1, even when it was 'in the money'. If $S_1 = 80$, we make 20 from exercising immediately, but there is 25 to be made from waiting.

2   In this example there was never a strictly positive advantage to early exercise of the option. It was always at least as good to wait. In fact if interest rates are zero this is *always* the case, as is shown in Exercise 6. For non-zero interest rates, early exercise can be optimal, see Exercise 9.

□

2.3   Discrete parameter martingales and Markov processes

Our multiperiod stock market model still looks rather special. To prepare the ground for the continuous time world of later chapters we now place it in the more general framework of discrete parameter martingales and Markov processes.

First we recall the concepts of random variables and stochastic processes.

Random
variables

Formally, when we talk about a random variable we must first specify a probability triple $(\Omega, \mathcal{F}, \mathbb{P})$, where Ω is a set, the *sample space*, $\mathcal{F}$ is a collection of subsets of Ω, *events*, and $\mathbb{P}$ specifies the probability of each event $A \in \mathcal{F}$. The collection $\mathcal{F}$ is a σ-field, that is, $\Omega \in \mathcal{F}$ and $\mathcal{F}$ is closed under the operations of countable union and taking complements. The probability $\mathbb{P}$ must satisfy the usual *axioms of probability*:

- $0 \leq \mathbb{P}[A] \leq 1$, for all $A \in \mathcal{F}$,
- $\mathbb{P}[\Omega] = 1$,
- $\mathbb{P}[A \cup B] = \mathbb{P}[A] + \mathbb{P}[B]$ for any disjoint $A, B \in \mathcal{F}$,
- if $A_n \in \mathcal{F}$ for all $n \in \mathbb{N}$ and $A_1 \subseteq A_2 \subseteq \cdots$ then $\mathbb{P}[A_n] \uparrow \mathbb{P}\left[\bigcup_n A_n\right]$ as $n \uparrow \infty$.

Definition 2.3.1   *A real-valued random variable, X, is a real-valued function on Ω that is $\mathcal{F}$-measurable. In the case of a discrete random variable (that is a random variable that can only take on countably many distinct values) this simply means*

$$\{\omega \in \Omega : X(\omega) = x\} \in \mathcal{F},$$

so that $\mathbb{P}$ assigns a probability to the event $\{X = x\}$. For a general real-valued random variable we require that

$$\{\omega \in \Omega : X(\omega) \leq x\} \in \mathcal{F},$$

so that we can define the distribution function, $F(x) = \mathbb{P}[X \leq x]$.

This looks like an excessively complicated way of talking about a relatively straightforward concept. It is technically required because it may not be *possible* to define $\mathbb{P}$ in a non-trivial way on *all* subsets of Ω, but most of the time we don't go far wrong if we ignore such technical details. However, when we start to study *stochastic processes*, random variables that evolve with time, it becomes much more natural to work in a slightly more formal framework.

Stochastic
processes

To specify a (discrete time) stochastic process, we typically require not just a single σ-field, $\mathcal{F}$, but an increasing sequence of them, $\mathcal{F}_n \subseteq \mathcal{F}_{n+1} \subseteq \cdots \subseteq \mathcal{F}$. The collection $\{\mathcal{F}_n\}_{n \geq 0}$ is then called a *filtration* and the quadruple $(\Omega, \mathcal{F}, \{\mathcal{F}_n\}_{n \geq 0}, \mathbb{P})$ is called a *filtered probability space*.

Definition 2.3.2   *A real-valued* stochastic process *is just a sequence of real-valued functions, $\{X_n\}_{n \geq 0}$, on Ω. We say that it is* adapted *to the filtration $\{\mathcal{F}_n\}_{n \geq 0}$ if X_n is $\mathcal{F}_n$-measurable for each n.*

One can then think of the σ-field $\mathcal{F}_n$ as encoding all the information about the evolution of the stochastic process up until time n. That is, if we know whether each event in $\mathcal{F}_n$ happens or not then we can infer the path followed by the stochastic process up until time n. We shall call the filtration that encodes *precisely* this information the *natural* filtration associated to the stochastic process $\{X_n\}_{n \geq 0}$.

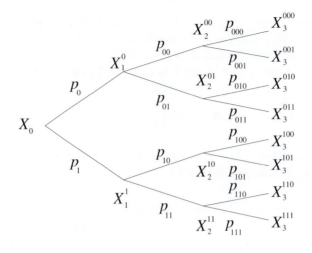

X_0

P_0 X_1^0 P_{00} X_2^{00} P_{000} X_3^{000}

P_{001} X_3^{001}

P_{01} X_2^{01} P_{010} X_3^{010}

P_{011} X_3^{011}

P_1 P_{10} P_{100} X_3^{100}

X_2^{10} P_{101} X_3^{101}

X_1^1 P_{110} X_3^{110}

P_{11} X_2^{11} P_{111} X_3^{111}

Figure 2.5   Tree representing the stochastic process of Example 2.3.3 and its distribution.

There is an important consequence of the very formal way that this is set up. Notice that we have defined the process $\{X_n\}_{n\geq 0}$ as a sequence of measurable functions on Ω *without reference to* $\mathbb{P}$. This is exactly analogous to the situation in our tree models. We specified the possible values that the stock price could take at time n, corresponding to prescribing the functions $\{X_n\}_{n\geq 0}$, and superposed the probabilities afterwards. Even if we had a preconception of what the probabilities of up and down jumps might be, we then *changed* probability (to the risk-neutral probabilities) in order actually to price claims. This process of changing probability will be fundamental to our approach to option pricing, even in our most complex market models.

Conditional expectation
When we constructed the probabilities on paths through our binary (or binomial) trees, we first specified the probability on each branch of the tree. This was done in such a way that the expected value of $e^{-r\delta t}S_{k+1}$ given that the value of the stock at time $k\delta t$ is known to be S_k is just S_k. This condition specifies the probabilities on the two branches emanating from the node corresponding to S_k at time $k\delta t$. We should like to extend this idea, but first we need to remind ourselves about *conditional expectation*. This is best explained through an example.

Example 2.3.3   *Consider the stochastic process represented by the tree in Figure 2.5. Its distribution is given by the probabilities on the branches of the tree, where, as in §2.1, we assume that the probability of a particular path through the tree is the product of the probabilities of the branches that comprise that path.*

Calculation: Our tree explicitly specifies $\{X_n\}_{n\geq 0}$ and, for a given Ω, implicitly specifies $\mathbb{P}$. In later examples we shall be less pedantic, but here we write down Ω explicitly. There are many possible choices, but an obvious one is the set of all possible sequences of 'up' and 'down' jumps. If $\omega = (u, u, d)$, say, then $X_1(\omega) = X_1^0$, $X_2(\omega) = X_2^{00}$ and $X_3(\omega) = X_3^{001}$.

First let us calculate the conditional expectation

$$\mathbb{E}[X_3|\mathcal{F}_1].$$

Using our interpretation of $\mathcal{F}_n$ as 'information up to time n', our problem is to determine the conditional expectation of X_3 given all the information up to time one. Notice that what we are calculating is an $\mathcal{F}_1$-measurable *random variable*. It depends only on what happened up until time one. There are just two possibilities: the first jump is up, or the first jump is down.

• If the first jump is up, the possible values of X_3 are X_3^{000}, X_3^{001}, X_3^{010} and X_3^{011}. The probability of each value is determined by the path probabilities but restricted to paths emanating from the upper node at time one. The conditional expectation then takes the value

$$\mathbb{E}[X_3|\mathcal{F}_1](u) = p_{00}p_{000}X_3^{000} + p_{00}p_{001}X_3^{001} + p_{01}p_{010}X_3^{010} + p_{01}p_{011}X_3^{011}.$$

This happens with probability p_0.

• If the first jump is down, which happens with probability p_1, the conditional expectation takes the value

$$\mathbb{E}[X_3|\mathcal{F}_1](d) = p_{10}p_{100}X_3^{100} + p_{10}p_{101}X_3^{101} + p_{11}p_{110}X_3^{110} + p_{11}p_{111}X_3^{111}.$$

Similarly, we can calculate $\mathbb{E}[X_3|\mathcal{F}_2]$. This random variable will be $\mathcal{F}_2$-measurable – its value depends on the first two jumps of the process. Its distribution is given in the table below.

Value	Probability
$\mathbb{E}[X|\mathcal{F}_2](uu) = p_{000}X_3^{000} + p_{001}X_3^{001}$	p_0p_{00}
$\mathbb{E}[X|\mathcal{F}_2](ud) = p_{010}X_3^{010} + p_{011}X_3^{011}$	p_0p_{01}
$\mathbb{E}[X|\mathcal{F}_2](du) = p_{100}X_3^{100} + p_{101}X_3^{101}$	p_1p_{10}
$\mathbb{E}[X|\mathcal{F}_2](dd) = p_{110}X_3^{110} + p_{111}X_3^{111}$	p_1p_{11}

Of course, since $\mathbb{E}[X_3|\mathcal{F}_2]$ is an $\mathcal{F}_2$-measurable random variable and $\mathcal{F}_1 \subseteq \mathcal{F}_2$, we can calculate the conditional expectation

$$\mathbb{E}[\mathbb{E}[X_3|\mathcal{F}_2]|\mathcal{F}_1].$$

$$\mathbb{E}[\mathbb{E}[X_3|\mathcal{F}_2]|\mathcal{F}_1](u) = p_{00}\mathbb{E}[X_3|\mathcal{F}_2](uu) + p_{01}\mathbb{E}[X_3|\mathcal{F}_2](ud),$$
$$\mathbb{E}[\mathbb{E}[X_3|\mathcal{F}_2]|\mathcal{F}_1](d) = p_{10}\mathbb{E}[X_3|\mathcal{F}_2](du) + p_{11}\mathbb{E}[X_3|\mathcal{F}_2](dd).$$

Substituting the value of $\mathbb{E}[X_3|\mathcal{F}_2]$ from the table above it is easily checked that this reduces to

$$\mathbb{E}[\mathbb{E}[X_3|\mathcal{F}_2]|\mathcal{F}_1] = \mathbb{E}[X_3|\mathcal{F}_1]. \tag{2.1}$$

□

Here is the formal definition.

Definition 2.3.4 (Conditional expectation)   *Suppose that X is an $\mathcal{F}$-measurable random variable with $\mathbb{E}[|X|] < \infty$. Suppose that $\mathcal{G} \subseteq \mathcal{F}$ is a σ-field; then the conditional expectation of X given $\mathcal{G}$, written $\mathbb{E}[X|\mathcal{G}]$, is the $\mathcal{G}$-measurable random variable with the property that for any $A \in \mathcal{G}$*

$$\mathbb{E}[[X|\mathcal{G}];A] \triangleq \int_A \mathbb{E}[X|\mathcal{G}]\,d\mathbb{P} = \int_A X\,d\mathbb{P} \triangleq \mathbb{E}[X;A].$$

The conditional expectation exists, but is only unique up to the addition of a random variable that is zero with probability one. This technical point will be important in Exercise 17 of Chapter 3.

Equation (2.1) is a special case of the following key property of conditional expectations.

The tower property of conditional expectations: Suppose that $\mathcal{F}_i \subseteq \mathcal{F}_j$; then

$$\mathbb{E}[\mathbb{E}[X|\mathcal{F}_j]|\mathcal{F}_i] = \mathbb{E}[X|\mathcal{F}_i].$$

In words this says that conditioning first on the information up to time j and then on the information up to an earlier time i is the same as conditioning originally up to time i.   □

In calculations with conditional expectations, it is often useful to remember the following fact.

Taking out what is known in conditional expectations: Suppose that $\mathbb{E}[X]$ and $\mathbb{E}[XY] < \infty$; then

$$\text{if } Y \text{ is } \mathcal{F}_n\text{-measurable,}\quad \mathbb{E}[XY|\mathcal{F}_n] = Y\mathbb{E}[X|\mathcal{F}_n].$$

This just says that if Y is known by time n, then if we condition on the information up to time n we can treat Y as constant.   □

The martingale property

The probability measure on the tree that we used in §2.1 to price claims was chosen so that if we define $\{\tilde{S}_k\}_{k\geq 0}$ to be the discounted stock price, that is $\tilde{S}_k = e^{-kr\delta t}S_k$, then the expected value of $\tilde{S}_{k+1}$ given that we know $\tilde{S}_k$ is just $\tilde{S}_k$. We use the notation

$$\mathbb{E}[\tilde{S}_{k+1}|\tilde{S}_k] = \tilde{S}_k.$$

Because in our model the stock price has 'no memory', so that the movement of the stock over the next tick of the clock is not influenced by the way in which it reached its current value, conditioning on knowing $\tilde{S}_k$ is actually the same as conditioning on knowing all of $\mathcal{F}_k$, so that

$$\mathbb{E}\big[\tilde{S}_{k+1}|\mathcal{F}_k\big] = \tilde{S}_k. \tag{2.2}$$

The property (2.2) is sufficiently important that it has a name.

Definition 2.3.5   *Suppose that $\big(\Omega, \{\mathcal{F}_n\}_{n\geq 0}, \mathcal{F}, \mathbb{P}\big)$ is a filtered probability space. The sequence of random variables $\{X_n\}_{n\geq 0}$ is a* martingale *with respect to $\mathbb{P}$ and $\{\mathcal{F}_n\}_{n\geq 0}$ if*

$$\mathbb{E}\left[|X_n|\right] < \infty, \quad \forall n, \tag{2.3}$$

and

$$\mathbb{E}\left[X_{n+1}|\mathcal{F}_n\right] = X_n, \quad \forall n. \tag{2.4}$$

If we replace equation (2.4) by

$$\mathbb{E}\left[X_{n+1}|\mathcal{F}_n\right] \leq X_n, \quad \forall n,$$

*then $\{X_n\}_{n\geq 0}$ is a $\big(\mathbb{P}, \{\mathcal{F}_n\}_{n\geq 0}\big)$-*super*martingale. If instead we replace it by*

$$\mathbb{E}\left[X_{n+1}|\mathcal{F}_n\right] \geq X_n, \quad \forall n,$$

*then $\{X_n\}_{n\geq 0}$ is a $\big(\mathbb{P}, \{\mathcal{F}_n\}_{n\geq 0}\big)$-*sub*martingale.*

These definitions are not exhaustive. There are plenty of processes that fall into none of these categories. A martingale is often thought of as tracking the net gain after successive plays of a fair game. In this setting a supermartingale models net gain from playing an unfavourable game (one we are more likely to lose than to win) and a submartingale is the net gain from playing a favourable game.

It is extremely important to note that the notion of a martingale is really that of a $\big(\mathbb{P}, \{\mathcal{F}_n\}_{n\geq 0}\big)$-martingale. Recall that our definition of stochastic process has divorced the rôles of the sequence $\{\mathcal{F}_n\}_{n\geq 0}$, the $\mathcal{F}$-measurable functions $\{X_n\}_{n\geq 0}$ on Ω and the probability measure $\mathbb{P}$ defined on elements of $\mathcal{F}$. In the setting of §2.1, our view of the market may be that the discounted stock price is *not* a martingale (indeed it probably isn't or no one would ever speculate on stocks – they could get the same money, risk-free, by buying cash bonds). We *change* the probability measure to one which makes the discounted stock price a martingale for the purposes of pricing and, as we shall see, hedging. We shall refer to the probability measure that represents our view of the market as the *market measure*. The new probability measure, which we use for pricing and hedging, is known as the *equivalent martingale measure*.

Remark: (*Martingales indexed by a subset of $\mathbb{N}$*) Although we have defined martingales indexed by $n \in \mathbb{N}$, we shall often talk about martingales indexed by $\{0 \leq n \leq N\}$. They are defined by restricting conditions (2.3) and (2.4) to $\{0 \leq n \leq N\}$. We shall state our key results for martingales indexed by $\{n \geq 0\}$; they can be modified in the obvious way to apply to martingales indexed by $\{0 \leq n \leq N\}$.   □

It is often useful to observe that, by the tower property, if $\{X_n\}_{n\geq0}$ is a $\left(\mathbb{P}, \{\mathcal{F}_n\}_{n\geq0}\right)$-martingale then for $i < j$,

$$\mathbb{E}\left[X_j \middle| \mathcal{F}_i\right] = X_i.$$

The Markov property

Calculations can also be simplified if our martingales have an additional property: the Markov property.

Definition 2.3.6 (Markov process)   *The stochastic process $\{X_n\}_{n\geq0}$ (with its natural filtration, $\{\mathcal{F}_n\}_{n\geq0}$) is a discrete time Markov process if*

$$\mathbb{P}\left[X_{n+1} \in B \middle| \mathcal{F}_n\right] = \mathbb{P}\left[X_{n+1} \in B \middle| X_n\right],$$

for all $B \in \mathcal{F}$.

In words this says that the probability that $X_{n+1} \in B$ given that we know the whole history of the process up to time n is the same as the probability that $X_{n+1} \in B$ given only the value of X_n. A Markov process has *no memory*. Many of our examples of martingales (and all our examples of market models) will also have the Markov property. However, not all martingales are Markov processes and not all Markov processes are martingales (see Exercise 11).

> **Notation:**   When we wish to emphasise that a filtration is 'generated by' the stochastic process $\{X_n\}_{n\geq0}$ we use the notation $\{\mathcal{F}_n^X\}_{n\geq0}$.
> Unless otherwise stated, $\{\mathcal{F}_n\}_{n\geq0}$ will always be understood to mean the natural filtration associated with the stochastic process under consideration.

It would be excessively pedantic always to insist upon an explicit specification of Ω and so, generally, we won't. We shall also use '$\{X_n\}_{n\geq0}$ is a $\mathbb{P}$-martingale' to mean $\{X_n\}_{n\geq0}$ is a $\left(\mathbb{P}, \{\mathcal{F}_n^X\}_{n\geq0}\right)$-martingale'.

Examples

Example 2.3.7 (Random walk)   *A one-dimensional simple random walk, $\{S_n\}_{n\geq0}$, is a Markov process such that $S_{n+1} = S_n + \xi_{n+1}$ where (for each n) $\xi_n \in \{-1, +1\}$ and, under $\mathbb{P}$, $\{\xi_n\}_{n\geq0}$ are independent identically distributed random variables. Thus*

$$\mathbb{P}\left[S_{n+1} = k + 1 \middle| S_n = k\right] = p, \quad \mathbb{P}\left[S_{n+1} = k - 1 \middle| S_n = k\right] = 1 - p,$$

where $p \in [0, 1]$.

If $p = 0.5$, then $\{S_n\}_{n\geq0}$ is a $\mathbb{P}$-martingale. If $p < 0.5$ (resp. $p > 0.5$), then $\{S_n\}_{n\geq0}$ is a $\mathbb{P}$-supermartingale (resp. $\mathbb{P}$-submartingale).

Justification: To check this, notice that since the random walk can be a distance at most n from its starting point at time n, the expectation $\mathbb{E}\left[|S_n|\right] < \infty$ is evidently finite. Moreover,

$$\begin{aligned}
\mathbb{E}\left[S_{n+1}|\,\mathcal{F}_n\right] &= \mathbb{E}\left[S_n + \xi_{n+1}|\,\mathcal{F}_n\right]\\
&= S_n + \mathbb{E}\left[\xi_{n+1}|\,\mathcal{F}_n\right]\\
&= S_n + \mathbb{E}\left[\xi_{n+1}\right],
\end{aligned}$$

where we have used independence of the $\{\xi_n\}_{n\geq 0}$ in the last line. It suffices then to observe that

$$\mathbb{E}\left[\xi_{n+1}\right]\begin{cases} < 0, & p < 0.5,\\ = 0, & p = 0.5,\\ > 0, & p > 0.5. \end{cases}$$

$\square$

Example 2.3.8 (Conditional expectation of a claim)   *Suppose that Ω and a filtration $\{\mathcal{F}_n\}_{n\geq 0}$ are given. (The example that we have in mind is that $\mathcal{F}_n$ encodes the history of a financial market up until time $n\delta t$.) Let C_N be any bounded $\mathcal{F}_N$-measurable random variable. (This we are thinking of as a claim against us at time $N\delta t$.) Then for* any *probability measure $\mathbb{P}$, the conditional expectation process, $\{X_n\}_{0\leq n\leq N}$, given by*

$$X_n = \mathbb{E}\left[C_N|\,\mathcal{F}_n\right],$$

is a $\left(\mathbb{P}, \{\mathcal{F}_n\}_{0\leq n\leq N}\right)$-martingale.

Example 2.3.9 (The discounted price of a claim)   *In solving our pricing problem for a European option with value C_N at the expiry time $N\delta t$ in the multiperiod binary model of stock prices of §2.1, we found a probability measure, which we denote by $\mathbb{Q}$, under which the discounted stock price is a martingale. For any claim, C_N, at time $N\delta t$, provided $\mathbb{E}^{\mathbb{Q}}\left[|C_N|\right] < \infty$, the fair price at time $n\delta t$ of an option with payoff C_N at time $N\delta t$ was found to be*

$$V_n = e^{-r(N-n)\delta t}\mathbb{E}^{\mathbb{Q}}\left[C_N|\,\mathcal{F}_n\right].$$

Define the discounted claim process by $\tilde{V}_n = e^{-rn\delta t}V_n$. Then $\{\tilde{V}_n\}_{0\leq n\leq N}$ is a $\mathbb{Q}$-martingale. This would remain true even if we dropped the assumption of constant interest rates, provided that we knew the risk-free rate over the time interval $[i\delta t, (i+1)\delta t)$ at the beginning *of the period.*

New martingales from old

Our last example shows that the discounted price process of a European option is a martingale. In other words, the discounted value of our replicating portfolio is a martingale. As before, we write (ϕ_n, ψ_n) for the amount of stock and bond held in the replicating portfolio over the nth time interval, that is $[(n-1)\delta t, n\delta t)$. The value of the portfolio at time $n\delta t$ is then

$$V_n = \phi_{n+1}S_n + \psi_{n+1}B_n,$$

where B_n is the value of the cash bond at time $n\delta t$. The portfolio is *self-financing*, that is the cost of constructing the new portfolio at time $(n+1)\delta t$ is exactly offset by the proceeds of selling the portfolio that we have held over $[n\delta t, (n + 1)\delta t)$. In symbols,

$$\phi_{n+1} S_{n+1} + \psi_{n+1} B_{n+1} = \phi_{n+2} S_{n+1} + \psi_{n+2} B_{n+1}.$$

The discounted price is

$$\tilde{V}_n = \phi_{n+1} \tilde{S}_n + \psi_{n+1},$$

and since, using the self-financing property,

$$\phi_{n+1} \tilde{S}_{n+1} + \psi_{n+1} = \phi_{n+2} \tilde{S}_{n+1} + \psi_{n+2},$$

we have

$$
\begin{aligned}
\tilde{V}_{n+1} - \tilde{V}_n &= \phi_{n+2} \tilde{S}_{n+1} + \psi_{n+2} - \phi_{n+1} \tilde{S}_n - \psi_{n+1} \\
&= \phi_{n+1} \left(\tilde{S}_{n+1} - \tilde{S}_n \right).
\end{aligned}
$$

That is

$$\tilde{V}_n = V_0 + \sum_{j=0}^{n-1} \phi_{j+1} \left(\tilde{S}_{j+1} - \tilde{S}_j \right). \tag{2.5}$$

From our earlier remarks, $\{\tilde{V}_n\}_{0 \leq n \leq N}$ is a $\mathbb{Q}$-martingale, so what we have checked is that under the probability measure $\mathbb{Q}$ for which $\{\tilde{S}_n\}_{0 \leq n \leq N}$ is a martingale, the expression on the right hand side of equation (2.5) is also a martingale. This is part of a general phenomenon. To state a precise result we need a definition. Recall that we knew ϕ_i at time $(i - 1)\delta t$.

Definition 2.3.10   *Given a filtration $\{\mathcal{F}_n\}_{n \geq 0}$, the process $\{A_n\}_{n \geq 1}$ is $\{\mathcal{F}_n\}_{n \geq 0}$-previsible or $\{\mathcal{F}_n\}_{n \geq 0}$-predictable if A_n is $\mathcal{F}_{n-1}$-measurable for all $n \geq 1$.*

Note that this is the sort of randomness that we have permitted for our cash bond.

Discrete
stochastic
integrals

Proposition 2.3.11   *Suppose that $\{X_n\}_{n \geq 0}$ is adapted to the filtration $\{\mathcal{F}_n\}_{n \geq 0}$ and that $\{\phi_n\}_{n \geq 1}$ is $\{\mathcal{F}_n\}_{n \geq 0}$-previsible. Define*

$$Z_n = Z_0 + \sum_{j=0}^{n-1} \phi_{j+1} \left(X_{j+1} - X_j \right), \tag{2.6}$$

where Z_0 is a constant.
    If $\{X_n\}_{n \geq 0}$ is a $\left(\mathbb{P}, \{\mathcal{F}_n\}_{n \geq 0} \right)$-martingale, then so is $\{Z_n\}_{n \geq 0}$.

Remark: If $\{\theta_n\}_{n \geq 0}$ is *adapted* to $\{\mathcal{F}_n\}_{n \geq 0}$, then the process $\{\phi_n\}_{n \geq 1}$ defined by $\phi_n = \theta_{n-1}$ is previsible. Thus for an $\{\mathcal{F}_n\}_{n \geq 0}$-adapted process $\{\theta_n\}_{n \geq 0}$, if $\{X_n\}_{n \geq 0}$ is a $\left(\mathbb{P}, \{\mathcal{F}_t\}_{t \geq 0} \right)$-martingale then so is

$$Z_n = Z_0 + \sum_{j=0}^{n-1} \theta_j \left(X_{j+1} - X_j \right).$$

□

Proof of Proposition 2.3.11: This is an exercise in the use of conditional expectations.

$$
\begin{aligned}
\mathbb{E}\left[Z_{n+1} \mid \mathcal{F}_n\right] - Z_n &= \mathbb{E}\left[Z_{n+1} - Z_n \mid \mathcal{F}_n\right] \\
&= \mathbb{E}\left[\phi_{n+1}\left(X_{n+1} - X_n\right) \mid \mathcal{F}_n\right] \\
&= \phi_{n+1}\mathbb{E}\left[\left(X_{n+1} - X_n\right) \mid \mathcal{F}_n\right] \\
&= \phi_{n+1}\left(\mathbb{E}\left[X_{n+1} \mid \mathcal{F}_n\right] - X_n\right) \\
&= 0.
\end{aligned}
$$

□

We can think of the sum in equation (2.6) as a *discrete stochastic integral*. When we turn to stochastic integration in Chapter 4, we shall essentially be passing to limits in sums of this form.

The Fundamental Theorem of Asset Pricing

It is not just our binomial models that can be incorporated into the martingale framework. The same argument that allows us to pass from the single period to the multiperiod binary model allows us to pass from the single period models of §1.5 and §1.6 to a multiperiod model. We now recast Theorems 1.5.2 and 1.6.2 in this language. Suppose that our market consists of K stocks and that the possible values that the stock prices $S^1, \ldots, S^K$ can take on at times $\delta t, 2\delta t, 3\delta t, \ldots, N\delta t = T$ are known. We denote by Ω the set of all possible 'paths' that the stock price vector can follow in $\mathbb{R}_+^K$.

Theorem 1.5.2 tells us that the absence of arbitrage is equivalent to the existence of a probability measure, $\mathbb{Q}$, on Ω that assigns strictly positive mass to every $\omega \in \Omega$ and such that

$$
S_{r-1} = \psi_0^{(r)}\mathbb{E}^{\mathbb{Q}}[S_r \mid S_{r-1}],
$$

where S_r is the vector of stock prices at time r and $\psi_0^{(r)}$ is the discount on riskless borrowing over $[(r-1)\delta t, r\delta t]$.

If, as above, we consider the *discounted* stock prices, $\{\tilde{S}_j\}_{0 \le j \le N}$, given by $\tilde{S}_j = \prod_{i=1}^{j} \psi_0^{(i)} S_j$, then

$$
\mathbb{E}^{\mathbb{Q}}[\tilde{S}_r \mid \tilde{S}_1, \ldots, \tilde{S}_{r-1}] = \mathbb{E}^{\mathbb{Q}}\left[\tilde{S}_r \mid \mathcal{F}_{r-1}\right] = \tilde{S}_{r-1}.
$$

In other words, the discounted stock price vector is a $\mathbb{Q}$-*martingale*.

Definition 2.3.12 *Two probability measures $\mathbb{P}$ and $\mathbb{Q}$ on a space Ω are said to be equivalent if for all events $A \subseteq \Omega$*

$$
\mathbb{Q}(A) = 0 \quad \textit{if and only if} \quad \mathbb{P}(A) = 0.
$$

Suppose then that we have a market model in which the stock price vector can follow one of a finite number of paths Ω through $\mathbb{R}_+^K$. We may even have our own belief as to how the price will evolve, encoded in a probability measure, $\mathbb{P}$, on Ω. Theorem 1.5.2 and Theorem 1.6.2 combine to say:

Theorem 2.3.13   *For the multiperiod market model described above, there is no arbitrage if and only if there is an equivalent martingale measure $\mathbb{Q}$. That is, there is a measure, $\mathbb{Q}$, equivalent to $\mathbb{P}$, such that the discounted stock price process is a $\mathbb{Q}$-martingale.*

In that case, the time zero market price of an attainable claim C_N (to be delivered at time $N\delta t$) is unique and is given by

$$\mathbb{E}^{\mathbb{Q}}[\psi_0 C_N],$$

where $\psi_o = \prod_1^N \psi_0^{(i)}$ is the discount factor over N periods.

Although there are extra technical conditions, this fundamental theorem has essentially the same statement for markets that evolve continuously with time.

2.4   Some important martingale theorems

Phrasing everything in the martingale framework places many powerful theorems at our disposal. In this section, we present some of the most important results in the theory of discrete parameter martingales. However, our coverage is necessarily cursory. An excellent and highly readable account is Williams (1991).

Stopping times

One of the most important calculational tools in martingale theory is the Optional Stopping Theorem. Before we can state it, we need to introduce the notion of a stopping time.

Definition 2.4.1   *Given a sample space Ω equipped with a filtration $\{\mathcal{F}_n\}_{n\geq 0}$, a stopping time or optional time is a random variable $T : \Omega \to \mathbb{Z}_+$ with the property that*

$$\{T \leq n\} \in \mathcal{F}_n, \qquad \text{for all } n \geq 0.$$

This just says that we can decide whether or not $T \leq n$ on the basis of the information available at time n – we don't need to look into the future.

Example 2.4.2   *Consider the simple random walk of Example 2.3.7. Define T to be the first time that the random walk takes the value 1, that is*

$$T = \inf\{i \geq 0 : S_i = 1\};$$

then T is a stopping time.
        On the other hand,

$$U = \sup\{i \geq 0 : S_i = 1\}$$

is not *a stopping time.*

Optional stopping

An equivalent definition of stopping time is that the random variable $\theta_n \triangleq \mathbf{1}_{\{T\geq n+1\}}$, for $n \geq 0$, is *adapted* (see Definition 2.3.2). Consequently, from the remark following

Proposition 2.3.11, if $\{X_n\}_{n\geq 0}$ is a martingale, then so is the process

$$Z_n \triangleq \sum_{j=0}^{n-1} \theta_j \left(X_{j+1} - X_j\right). \tag{2.7}$$

Notice that we can rearrange this expression,

$$
\begin{aligned}
Z_n &= \sum_{j=0}^{n-1} \theta_j \left(X_{j+1} - X_j\right) \\
&= \sum_{j=0}^{n-1} \mathbf{1}_{\{T \geq j+1\}} \left(X_{j+1} - X_j\right) \\
&= X_{T \wedge n} - X_0,
\end{aligned}
$$

where $T \wedge n$ denotes the minimum of T and n.

Theorem 2.4.3 (Optional Stopping Theorem)   *Let $\left(\Omega, \mathcal{F}, \{\mathcal{F}_n\}_{n\geq 0}, \mathbb{P}\right)$ be a filtered probability space. Suppose that the process $\{X_n\}_{n\geq 0}$ is a $\left(\mathbb{P}, \{\mathcal{F}_n\}_{n\geq 0}\right)$-martingale, and that T is a bounded stopping time. Then*

$$\mathbb{E}[X_T | \mathcal{F}_0] = X_0,$$

and hence

$$\mathbb{E}[X_T] = X_0.$$

Proof: The proof is a simple application of the calculation that we did above. If we know that $T \leq N$, then in the notation of (2.7), $Z_N = X_T - X_0$ and since $\{Z_n\}_{n\geq 0}$ is a martingale, $\mathbb{E}[Z_N | \mathcal{F}_0] = Z_0 = 0$, i.e.

$$\mathbb{E}[X_T | \mathcal{F}_0] = X_0.$$

Taking expectations once again yields

$$\mathbb{E}[X_T] = X_0.$$

$\square$

It is essential in this result that the stopping time be *bounded*. In practice this will be the case in all of our financial applications, but Exercise 15 shows what can go wrong. More general versions of the theorem are available; see for example, Williams (1991). Here we satisfy ourselves with an application (see also Exercise 14).

Proposition 2.4.4   *Let $\{S_n\}_{n\geq 0}$ be the (asymmetric) simple random walk of Example 2.3.7 with $p > 1/2$. For $x \in \mathbb{Z}$ we write*

$$T_x = \inf\{n : S_n = x\},$$

and define

$$\phi(x) = \left(\frac{1-p}{p}\right)^x.$$

Then for $a < 0 < b$,

$$\mathbb{P}[T_a < T_b] = \frac{1 - \phi(b)}{\phi(a) - \phi(b)}.$$

Proof: We first show that $\{\phi(S_n)\}_{n \geq 0}$ is a $\mathbb{P}$-martingale. Since the walk can only take one step at a time, $-n \leq S_n \leq n$. Using also that $0 < (1-p)/p < 1$ for $p > 1/2$, we evidently have that

$$\mathbb{E}[|\phi(S_n)|] < \infty, \qquad \forall n.$$

To check that we really have a martingale is reduced to another exercise in conditional expectations. We must calculate

$$\mathbb{E}[\phi(S_{n+1})|\mathcal{F}_n].$$

Recall that $S_{n+1} = \sum_{j=1}^{n+1} \xi_j = S_n + \xi_{n+1}$, where, under $\mathbb{P}$, the random variables ξ_j are independent and identically distributed with

$$\mathbb{P}[\xi_j = 1] = p \quad \text{and} \quad \mathbb{P}[\xi_j = -1] = 1 - p.$$

This gives

$$
\begin{aligned}
\mathbb{E}[\phi(S_{n+1})|\mathcal{F}_n] &= \mathbb{E}\left[\phi(S_n)\left(\frac{1-p}{p}\right)^{\xi_{n+1}} \middle| \mathcal{F}_n\right] \\
&= \phi(S_n)\mathbb{E}\left[\left(\frac{1-p}{p}\right)^{\xi_{n+1}}\right] \\
&= \phi(S_n)\left(p\left(\frac{1-p}{p}\right)^1 + (1-p)\left(\frac{1-p}{p}\right)^{-1}\right) \\
&= \phi(S_n).
\end{aligned}
$$

We should now like to apply the Optional Stopping Theorem to the stopping time $T = T_a \wedge T_b$, the first time that the walk hits either a or b. The difficulty is that T is not *bounded*. Instead then, we apply the theorem to the stopping time $T \wedge N$ for an arbitrary (deterministic) N. This gives

$$
\begin{aligned}
1 &= \mathbb{E}[\phi(S_0)] = \mathbb{E}[\phi(S_{T \wedge N})] \\
&= \phi(a)\mathbb{P}[S_T = a, T \leq N] + \phi(b)\mathbb{P}[S_T = b, T \leq N] + \mathbb{E}[\phi(S_N), T > N].
\end{aligned}
$$

$$(2.8)$$

Now

$$
\begin{aligned}
0 \leq \mathbb{E}[\phi(S_N), T > N] &= \mathbb{E}[\phi(S_N)| T > N]\mathbb{P}[T > N] \\
&\leq \left[\left(\frac{1-p}{p}\right)^b + \left(\frac{p}{1-p}\right)^a\right]\mathbb{P}[T > N],
\end{aligned}
$$

and since $\mathbb{P}[T > N] \to 0$ as $N \to \infty$, we can let $N \to \infty$ in (2.8) to deduce that

$$\phi(a)\mathbb{P}[S_T = a] + \phi(b)\mathbb{P}[S_T = b] = 1. \tag{2.9}$$

Finally, since $\mathbb{P}[S_T = a] = 1 - \mathbb{P}[S_T = b]$, and $\mathbb{P}[T_a < T_b] = \mathbb{P}[S_T = a]$, equation (2.9) becomes

$$\phi(a)\mathbb{P}[T_a < T_b] + \phi(b)(1 - \mathbb{P}[T_a < T_b]) = 1.$$

Rearranging,

$$\mathbb{P}[T_a < T_b] = \frac{1 - \phi(b)}{\phi(a) - \phi(b)},$$

as required.   □

A
convergence
theorem

Often one can deduce a great deal about martingales from apparently scant information. An example is the result of Exercise 12 which says that a previsible martingale
is constant. Another example is provided by the following result.

Theorem 2.4.5 (Positive Supermartingale Convergence Theorem)   *If $\{X_n\}_{n\geq 0}$ is a*
$(\mathbb{P}, \{\mathcal{F}_n\}_{n\geq 0})$-supermartingale and $X_n \geq 0$ for all n, then there exists an $\mathcal{F}_\infty$-
measurable random variable, X_∞, with $\mathbb{E}[X_\infty] < \infty$ such that with $\mathbb{P}$-probability
one

$$X_n \to X_\infty \quad as \; n \to \infty.$$

A proof of this result is beyond our scope here, but can be found, for example, in
Williams (1991).

Compensation

Before returning to some finance, we record just one more result. Recall that
submartingales tend to rise on the average and supermartingales fall on the average.
The following result, sometimes called *compensation*, says that we can subtract a
non-decreasing process from a submartingale to obtain a martingale and we can add
a non-decreasing process to a supermartingale to obtain a martingale. In both cases,
the interesting thing is that the non-decreasing processes are *previsible*.

Proposition 2.4.6

1   *Suppose that $\{X_n\}_{n\geq 0}$ is a $(\mathbb{P}, \{\mathcal{F}_n\}_{n\geq 0})$-submartingale. Then there is a previs-*
    ible, non-decreasing process $\{A_n\}_{n\geq 0}$ such that $\{X_n - A_n\}_{n\geq 0}$ is a $(\mathbb{P}, \{\mathcal{F}_n\}_{n\geq 0})$-
    martingale. If we insist that $A_0 = 0$, then $\{A_n\}_{n\geq 0}$ is unique.
2   *Suppose that $\{X_n\}_{n\geq 0}$ is a $(\mathbb{P}, \{\mathcal{F}_n\}_{n\geq 0})$-supermartingale. Then there is a previs-*
    ible, non-decreasing process $\{A_n\}_{n\geq 1}$ such that $\{X_n + A_n\}_{n\geq 0}$ is a $(\mathbb{P}, \{\mathcal{F}_n\}_{n\geq 0})$-
    martingale. If we insist that $A_0 = 0$, then $\{A_n\}_{n\geq 0}$ is unique.

Proof: The proofs of the two parts are essentially identical, so we restrict our
attention to 1.

Define $A_0 = 0$ and then for $n \geq 1$ set

$$A_n - A_{n-1} = \mathbb{E}\left[X_n - X_{n-1} \mid \mathcal{F}_{n-1}\right].$$

By definition $\{A_n\}_{n\geq 0}$ will be previsible and non-decreasing (since $\{X_n\}_{n\geq 0}$ is a submartingale). We must check that $\{X_n - A_n\}_{n\geq 0}$ is a martingale. First we check that $\mathbb{E}\left[|X_n - A_n|\right] < \infty$ for all n.

$$
\begin{aligned}
\mathbb{E}\left[|X_n - A_n|\right] &\leq \mathbb{E}\left[|X_n|\right] + \mathbb{E}\left[A_n\right] \\
&= \mathbb{E}\left[|X_n|\right] + \mathbb{E}\left[A_0 + \sum_{j=1}^{n}\left(A_j - A_{j-1}\right)\right] \\
&= \mathbb{E}\left[|X_n|\right] + \sum_{j=1}^{n}\mathbb{E}\left[\mathbb{E}\left[X_j - X_{j-1} \mid \mathcal{F}_{j-1}\right]\right] \quad \text{(by definition of } A_j) \\
&\leq \mathbb{E}\left[|X_n|\right] + \sum_{j=1}^{n}\mathbb{E}\left[\mathbb{E}\left[|X_j| + |X_{j-1}| \mid \mathcal{F}_{j-1}\right]\right] \\
&= \mathbb{E}\left[|X_n|\right] + \sum_{j=1}^{n}\mathbb{E}\left[|X_j| + |X_{j-1}|\right] \quad \text{(tower property)},
\end{aligned}
$$

and evidently this final expression is finite since by assumption $\mathbb{E}\left[|X_j|\right] < \infty$ for all j.

Now we check the martingale property,

$$
\begin{aligned}
\mathbb{E}\left[X_{n+1} - A_{n+1} \mid \mathcal{F}_n\right] \\
&= \mathbb{E}\left[X_{n+1} - \mathbb{E}\left[X_{n+1} - X_n \mid \mathcal{F}_n\right] - A_n \mid \mathcal{F}_n\right] \quad \text{(by definition of } A_{n+1}) \\
&= \mathbb{E}\left[X_{n+1} - X_{n+1} + X_n - A_n \mid \mathcal{F}_n\right] \\
&= X_n - A_n.
\end{aligned}
$$

It remains to check that if $A_0 = 0$ then the process $\{A_n\}_{n\geq 0}$ is unique. Suppose that there were another predictable process $\{B_n\}_{n\geq 0}$ with the same property. Then $\{X_n - A_n\}_{n\geq 0}$ and $\{X_n - B_n\}_{n\geq 0}$ are both martingales and, therefore, so is the difference between them, $\{A_n - B_n\}_{n\geq 0}$. On the other hand $\{A_n - B_n\}_{n\geq 0}$ is predictable and *predictable martingales are constant* (see Exercise 12). Since $A_0 = 0 = B_0$, the proof is complete.   □

Let's see what these concepts correspond to in a financial example.

Example 2.4.7 (American options revisited)   *Assume the binomial model and notation of §2.2 and let $\mathbb{Q}$ be the probability measure on the tree under which the discounted stock price $\{\tilde{S}_n\}_{0\leq n\leq N}$ is a martingale. We denote by $\{\tilde{V}_n\}_{0\leq n\leq N}$ the discounted value of an American call or put option with strike K and maturity $T = N\delta t$ and define*

$$
\tilde{B}_n = \begin{cases}
e^{-n\delta t}(S_n - K)_+ & \text{in the case of the call,} \\
e^{-n\delta t}(K - S_n)_+ & \text{in the case of the put.}
\end{cases}
$$

(The filtration is always that generated by $\{S_n\}_{0 \le n \le N}$.) Then $\{\tilde{V}_n\}_{0 \le n \le N}$ is the smallest $\mathbb{Q}$-supermartingale that dominates $\{\tilde{B}_n\}_{0 \le n \le N}$.

In Exercise 16 it is shown that this characterisation provides yet another simple proof of Lemma 2.2.2.

Explanation for example: We know from §2.2 that

$$\tilde{V}_{n-1} = \max\left\{\tilde{B}_{n-1}, \mathbb{E}^{\mathbb{Q}}\left[\tilde{V}_n \,\middle|\, \mathcal{F}_{n-1}\right]\right\}, \quad 0 \le n \le N,$$

and $\tilde{V}_N = \tilde{B}_N$. Evidently $\{\tilde{V}_n\}_{0 \le n \le N}$ is a supermartingale that dominates $\{\tilde{B}_n\}_{0 \le n \le N}$. To check that it is the *smallest* supermartingale with this property, suppose that $\{\tilde{U}_n\}_{0 \le n \le N}$ is any other supermartingale that dominates $\{\tilde{B}_n\}_{n \ge 0}$. Then $\tilde{U}_N \ge \tilde{V}_N$, and if $\tilde{U}_n \ge \tilde{V}_n$, then

$$\tilde{U}_{n-1} \ge \mathbb{E}^{\mathbb{Q}}\left[\tilde{U}_n \,\middle|\, \mathcal{F}_{n-1}\right] \ge \mathbb{E}^{\mathbb{Q}}\left[\tilde{V}_n \,\middle|\, \mathcal{F}_{n-1}\right],$$

and so

$$\tilde{U}_{n-1} \ge \max\left\{\tilde{B}_{n-1}, \mathbb{E}^{\mathbb{Q}}\left[\tilde{V}_n \,\middle|\, \mathcal{F}_{n-1}\right]\right\} = \tilde{V}_{n-1}.$$

The result follows by backwards induction. The process $\{\tilde{V}_n\}_{0 \le n \le N}$ is called the *Snell envelope* of $\{\tilde{B}_n\}_{0 \le n \le N}$.   □

Remark: Proposition 2.4.6 tells us that we can write

$$\tilde{V}_n = \tilde{M}_n - \tilde{A}_n$$

where $\{\tilde{M}_n\}_{n \ge 0}$ is a martingale and $\{\tilde{A}_n\}_{n \ge 0}$ is a non-decreasing process, with $A_0 = 0$. Since the market is complete, we can hedge M_N exactly by holding a portfolio that consists over the nth time step of ϕ_n units of stock and ψ_n units of cash bond. The seller of the American option would more than meet her liability by holding such a portfolio. The holder of the option will exercise at the first time j when $\tilde{A}_{j+1}$ is non-zero (recall that the process $\{\tilde{A}_n\}_{n \ge 0}$ is *previsible*), since at that time it is better to sell the option and invest the money according to the hedging portfolio $\{(\phi_n, \psi_n)\}_{j \le n \le N}$.   □

2.5   The Binomial Representation Theorem

Pricing a derivative in the martingale framework corresponds to taking an expectation. But arbitrage prices are only meaningful if we can construct a hedging portfolio. If we know the hedging portfolio then we saw in the discussion preceding Definition 2.3.10 that we can express the discounted value of the portfolio, and therefore of the derivative, as a 'discrete stochastic integral' of the stock holding in the portfolio with respect to the discounted stock price. In order to pass from the discounted price of the derivative to a hedging portfolio we need the following converse to Proposition 2.3.11. We work in the context of our binomial model of stock prices.

Theorem 2.5.1 (Binomial Representation Theorem)   *Suppose that the measure $\mathbb{Q}$ is such that the discounted binomial price process $\{\tilde{S}_n\}_{n\geq0}$ is a $\mathbb{Q}$-martingale. If $\{\tilde{V}_n\}_{n\geq0}$ is any other $(\mathbb{Q}, \{\mathcal{F}_n\}_{n\geq0})$-martingale, then there exists an $\{\mathcal{F}_n\}_{n\geq0}$-predictable process $\{\phi_n\}_{n\geq1}$ such that*

$$\tilde{V}_n = \tilde{V}_0 + \sum_{j=0}^{n-1} \phi_{j+1}\left(\tilde{S}_{j+1} - \tilde{S}_j\right). \tag{2.10}$$

Proof: We consider a single time step for our binomial tree. It is convenient to write

$$\Delta\tilde{V}_{i+1} = \tilde{V}_{i+1} - \tilde{V}_i \quad \text{and} \quad \Delta\tilde{S}_{i+1} = \tilde{S}_{i+1} - \tilde{S}_i.$$

Given their values at time $i\delta t$, each of $\tilde{V}_{i+1}$ and $\tilde{S}_{i+1}$ can take on one of two possible values that we denote by $\{\tilde{V}_{i+1}(u), \tilde{V}_{i+1}(d)\}$ and $\{\tilde{S}_{i+1}(u), \tilde{S}_{i+1}(d)\}$ respectively.

We should like to write $\Delta\tilde{V}_{i+1} = \phi_{i+1}\Delta\tilde{S}_{i+1} + k_{i+1}$, where ϕ_{i+1} and k_{i+1} are both known at time $i\delta t$. In other words we seek ϕ_{i+1} and k_{i+1} such that

$$\tilde{V}_{i+1}(u) - \tilde{V}_i = \phi_{i+1}\left(\tilde{S}_{i+1}(u) - \tilde{S}_i\right) + k_{i+1},$$

and

$$\tilde{V}_{i+1}(d) - \tilde{V}_i = \phi_{i+1}\left(\tilde{S}_{i+1}(d) - \tilde{S}_i\right) + k_{i+1}.$$

Solving this gives

$$\phi_{i+1} = \frac{\tilde{V}_{i+1}(u) - \tilde{V}_{i+1}(d)}{\tilde{S}_{i+1}(u) - \tilde{S}_{i+1}(d)}$$

and $k_{i+1} = \tilde{V}_{i+1}(u) - \tilde{V}_i - \phi_{i+1}\left(\tilde{S}_{i+1}(u) - \tilde{S}_i\right)$, both of which are known at time $i\delta t$.

Now $\{\tilde{V}_i\}_{i\geq0}$ and $\{\tilde{S}_i\}_{i\geq0}$ are both martingales so that

$$\mathbb{E}\left[\Delta\tilde{V}_{i+1} \,\middle|\, \mathcal{F}_i\right] = 0 = \mathbb{E}\left[\Delta\tilde{S}_{i+1} \,\middle|\, \mathcal{F}_i\right]$$

from which it follows that $k_{i+1} = 0$.
 In other words,

$$\Delta\tilde{V}_{i+1} = \phi_{i+1}\Delta\tilde{S}_{i+1},$$

where ϕ_{i+1} is known at time $i\delta t$. Induction ties together all these increments into the result that we want.   □

From martingale representation to replicating portfolio

From our previous work, we know that if $\{\tilde{V}_i\}_{i\geq0}$ is the discounted price of a claim, then such a predictable process $\{\phi_i\}_{i\geq1}$ arises as the stock holding when we construct our replicating portfolio. We should like to go the other way. Given $\{\phi_i\}_{i\geq1}$, can we construct a self-financing replicating portfolio? Not surprisingly, the answer is yes.

Construction strategy: At time i, buy a portfolio that consists of ϕ_{i+1} units of stock and $\tilde{V}_i - \phi_{i+1}\tilde{S}_i$ units of cash bond.

We must check that this strategy really works. It is convenient to write B_i for the value of the bond at time $i\delta t$.

Suppose that at time $i\delta t$ we have bought ϕ_{i+1} units of stock and $\left(\tilde{V}_i - \phi_{i+1}\frac{S_i}{B_i}\right)$ units of cash bond. This will cost us

$$\phi_{i+1}S_i + \left(\tilde{V}_i - \phi_{i+1}\frac{S_i}{B_i}\right)B_i = \tilde{V}_i B_i = V_i.$$

The value of this portfolio at time $(i+1)\delta t$ is then

$$
\begin{aligned}
\phi_{i+1}S_{i+1} + \left(\tilde{V}_i - \phi_{i+1}\frac{S_i}{B_i}\right)B_{i+1} &= B_{i+1}\left(\phi_{i+1}\left(\frac{S_{i+1}}{B_{i+1}} - \frac{S_i}{B_i}\right) + \tilde{V}_i\right) \\
&= \tilde{V}_{i+1}B_{i+1} \quad \text{(by the binomial representation)} \\
&= V_{i+1},
\end{aligned}
$$

which is exactly enough to construct our new portfolio at time $(i+1)\delta t$. Moreover, at time $N\delta t$ we have precisely the right amount of money to meet the claim against us.

Three steps to replication: There are three steps to pricing and hedging a claim C_T against us at time T.

- Find a probability measure $\mathbb{Q}$ under which the discounted stock price (with its natural filtration) is a martingale.
- Form the discounted value process,

$$\tilde{V}_i = e^{-ri\delta t}V_i = \mathbb{E}^{\mathbb{Q}}\left[e^{-rT}C_T \,\middle|\, \mathcal{F}_i\right].$$

- Find a predictable process $\{\phi_i\}_{1 \le i \le N}$ such that

$$\Delta\tilde{V}_i = \phi_i \Delta\tilde{S}_i.$$

2.6 Overture to continuous models

Before rigorously deriving the acclaimed Black–Scholes pricing formula for the value of a European option, we are going to develop a substantial body of material. As an appetiser though, we can use our discrete techniques to see what form our results must take in the continuous world.

It is easy to believe that we should be able to use a discrete model with very small time periods to approximate a continuous model. The Black–Scholes model is based on the lognormal model that we mentioned in §1.2. With this in mind, we choose our approximation to have constant growth rate and constant 'noise'.

Model with
constant
stock growth
and noise

The model is parametrised by the time period, δt, and three fixed constant parameters, v, σ and the riskless rate r.

• The cash bond has the form $B_t = e^{rt}$, which does not depend on the interval
size.

• The stock price process follows the nodes of a binomial tree. If the current value
of the stock is s, then over the next time period it moves to the new value

$$\begin{cases} s \exp\left(v\delta t + \sigma\sqrt{\delta t}\right) & \text{if up,} \\ s \exp\left(v\delta t - \sigma\sqrt{\delta t}\right) & \text{if down.} \end{cases}$$

Suppose our belief is that the jumps are equally likely to be up or down. So under
the *market* measure, $\mathbb{P}\left[\text{up jump}\right] = 1/2 = \mathbb{P}\left[\text{down jump}\right]$ at each time step.

For a fixed time t, set N to be the number of time periods until time t, that is
$N = t/\delta t$. Then

$$S_t = S_0 \exp\left(vt + \sigma\sqrt{t}\left(\frac{2X_N - N}{\sqrt{N}}\right)\right),$$

where X_N is the total number of the N separate jumps which were up jumps. To see
what happens as $\delta t \to 0$ (or equivalently $N \to \infty$) we call on the Central Limit
Theorem.

Theorem 2.6.1 (Central Limit Theorem)   *Let $\xi_1, \xi_2, \ldots$ be a sequence of independent identically distributed random variables under the probability measure $\mathbb{P}$ with
finite mean μ and finite non-zero variance σ^2 and let $S_n = \xi_1 + \ldots + \xi_n$. Then*

$$\frac{S_n - n\mu}{\sqrt{n\sigma^2}}$$

converges in distribution to an $N(0, 1)$ random variable as $n \to \infty$.

Now X_N is the sum of N independent random variables $\{\xi_i\}_{1 \le i \le N}$ taking the value
$+1$ with probability $\frac{1}{2}$ and 0 otherwise. This means $\mathbb{E}\left[\xi_i\right] = \frac{1}{2}$ and $var\left[\xi_i\right] = \frac{1}{4}$ so
that by the Central Limit Theorem, the distribution of the random variable $(2X_N - N)/\sqrt{N}$ converges to that of a normal random variable with mean zero and variance
one. In other words, as δt gets smaller (and so N gets larger), the distribution of S_t
converges to that of a lognormal distribution. More precisely, in the limit, $\log S_t$ is
normally distributed with mean $\log S_0 + vt$ and variance $\sigma^2 t$.

Under the
martingale
measure

This is what happens under the original measure $\mathbb{P}$. What happens under the
martingale measure, $\mathbb{Q}$, that we use for pricing?

By Lemma 1.3.2, under the martingale measure, the probability of an up jump is

$$p = \frac{\exp(r\delta t) - \exp(v\delta t - \sigma\sqrt{\delta t})}{\exp(v\delta t + \sigma\sqrt{\delta t}) - \exp(v\delta t - \sigma\sqrt{\delta t})},$$

which is approximately

$$\frac{1}{2}\left(1 - \sqrt{\delta t}\left(\frac{v + \frac{1}{2}\sigma^2 - r}{\sigma}\right)\right).$$

So under the martingale measure, $\mathbb{Q}$, X_N is still binomially distributed, but now has mean Np and variance $Np(1-p)$.

Thus, under $\mathbb{Q}$, $(2X_N - N)/\sqrt{N}$ has mean that tends to $-\sqrt{t}\left(v + \frac{1}{2}\sigma^2 - r\right)/\sigma$ and variance that approaches one as δt tends to zero. Again using the Central Limit Theorem the random variable $(2X_N - N)/\sqrt{N}$ converges to a normally distributed random variable, with mean $-\sqrt{t}\left(v + \frac{1}{2}\sigma^2 - r\right)/\sigma$ and variance one. Under $\mathbb{Q}$ then, S_t is lognormally distributed with mean $\log S_0 + (r - \frac{1}{2}\sigma^2)t$ and variance $\sigma^2 t$. This can be written

$$S_t = \exp\left(\sigma\sqrt{t}Z + \left(r - \frac{1}{2}\sigma^2\right)t\right),$$

where, under $\mathbb{Q}$, the random variable Z is normally distributed with mean zero and variance one.

Pricing a call option

If our discrete theory carries over to the continuous limit, then in our continuous model the price at time zero of a European call option with strike price K at time T will be the discounted expected value of the claim under the martingale measure, that is

$$\mathbb{E}^{\mathbb{Q}}\left[e^{-rT}(S_T - K)_+\right],$$

where r is the riskless rate. Substituting, we obtain

$$\mathbb{E}^{\mathbb{Q}}\left[\left(S_0 \exp\left(\sigma\sqrt{T}Z - \frac{1}{2}\sigma^2 T\right) - K\exp\left(-rT\right)\right)_+\right]. \qquad (2.11)$$

We'll derive this pricing formula rigorously in Chapter 5 where we'll also show that equation (2.11) can be evaluated as

$$S_0\Phi\left(\frac{\log\frac{S_0}{K} + \left(r + \frac{1}{2}\sigma^2\right)T}{\sigma\sqrt{T}}\right) - Ke^{-rT}\Phi\left(\frac{\log\frac{S_0}{K} + \left(r - \frac{1}{2}\sigma^2\right)T}{\sigma\sqrt{T}}\right),$$

where Φ is the standard normal distribution function,

$$\Phi(z) = \mathbb{Q}[Z \le z] = \int_{-\infty}^{z}\frac{1}{\sqrt{2\pi}}e^{-x^2/2}dx.$$

Exercises

1   Notice that, like the single period ternary model of Chapter 1, the two-step binomial model allows the stock to take on three distinct values at time 2. Show, however, that every claim can be exactly replicated by a self-financing portfolio, that is, the market is *complete*.

More generally, show that if the market evolves according to a k-step binomial model then it is complete.

2   Show that the price of a claim obtained by backwards induction on the binomial tree is precisely the value obtained by calculating the discounted expected value of the claim with respect to the path probabilities introduced in §2.1.

3   Consider two dates T_0, T_1 with $T_0 < T_1$. A *forward start option* is a contract in which the holder receives at time T_0, at no extra cost, an option with expiry date T_1 and strike price equal to S_{T_0} (the asset price at time T_0). Assume that the stock price evolves according to a two-period binary model, in which the asset price at time T_0 is either $S_0 u$ or $S_0 d$, and at time T_1 is one of $S_0 u^2$, $S_0 u d$ and $S_0 d^2$ with

$$d < \min \left\{ e^{rT_0}, e^{r(T_1 - T_0)} \right\} \leq \max \left\{ e^{rT_0}, e^{r(T_1 - T_0)} \right\} < u,$$

where r denotes the risk-free interest rate. Find the fair price of such an option at time zero.

4   A *digital option* is one in which the payoff depends in a discontinuous way on the asset price. The simplest example is the *cash-or-nothing option*, in which the payoff to the holder at maturity T is $X \mathbf{1}_{\{S_T > K\}}$ where X is some prespecified cash sum. Suppose that an asset price evolves according to the binomial model in which, at each step, the asset price moves from its current value S_n to one of $S_n u$ and $S_n d$. As usual, if ΔT denotes the length of each time step, $d < e^{r \Delta T} < u$.
Find the time zero price of the above option. You may leave your answer as a sum.

5   Let C_t denote the value at time t of an American call option on non-dividend-paying stock with strike price K and maturity T. If the risk-free interest rate is $r > 0$, prove that

$$C_t \geq S_t - K e^{-r(T-t)} > S_t - K,$$

and deduce that it is never optimal to exercise this option prior to the maturity time, T.

6   Let C_t be as in Exercise 5 and let P_t be the value of an American put option on the same stock with the same strike price and maturity. By comparing the values of two suitable portfolios, show that

$$C_t + K \geq P_t + S_t.$$

Using put–call parity for European options and the result of Exercise 5, show that

$$P_t \geq C_t + K e^{-r(T-t)} - S_t.$$

Combine these results to see that, if $r > 0$ and $t < T$,

$$S_t - K \leq C_t - P_t < S_t - K e^{-r(T-t)}$$

and deduce that if interest rates are zero, there is no advantage to early exercise of the put.

7 If a stock price is S just before a dividend D is paid, what is its value immediately after the payment? Suppose that a stock pays dividends at discrete times,
   $T_0, T_1, \dots, T_n$. Show that it can be optimal to exercise an American call on such a
   stock *prior* to expiry.

8 Suppose that the stock in Figure 2.3 will pay a dividend of 5% of its value at time
   2. As before, interest rates are zero and between times 2 and 3 the value of the stock
   will either increase or decrease by 20. Find the time zero price of an American call
   option on this stock with strike 100 and maturity 3. Is it ever optimal to exercise
   early?

9 Consider the American put option of Example 2.2.3, but now suppose that interest
   rates are such that a $1 cash bond at time $i\delta t$ is worth $1.1 at time $(i+1)\delta t$. Find the
   value of the put. At what time will it be exercised?

10 Suppose that an asset price evolves according to the binomial model. For simplicity
   suppose that the risk-free interest rate is zero and ΔT is 1. Suppose that under the
   probability $\mathbb{P}$, at each time step, stock prices go up with probability p and down with
   probability $1 - p$.
   The conditional expectation

$$M_n \triangleq \mathbb{E}[S_N | \mathcal{F}_n], \qquad 1 \le n \le N,$$

   is a stochastic process. Check that it is a $\mathbb{P}$-martingale and find the distribution of the
   random variable M_n.

11 (a) Find a Markov process that is not a martingale.
   (b) Find a martingale that is not a Markov process.

12 Show that a previsible martingale is constant.

13 Let $\{S_n\}_{n \ge 0}$ be simple random walk under the measure $\mathbb{P}$. Calculate $\mathbb{E}[S_n]$ and
   $var[S_n]$.

14 Let $\{S_n\}_{n \ge 0}$ be a symmetric simple random walk under the measure $\mathbb{P}$, that is, in the
   notation of Example 2.3.7, $p = 1/2$. Show that $\{S_n^2\}_{n \ge 0}$ is a $\mathbb{P}$-submartingale and
   that $\{S_n^2 - n\}_{n \ge 0}$ is a $\mathbb{P}$-martingale.
   Let $T = \inf\{n : S_n \notin (-a, a)\}$, where $a \in \mathbb{N}$. Use the Optional Stopping Theorem
   (applied to a suitable sequence of bounded stopping times) to show that $\mathbb{E}[T] = a^2$.

15 As in Exercise 14, let $\{S_n\}_{n \ge 0}$ be a symmetric simple random walk under $\mathbb{P}$ and write
   $X_n = S_n + 1$. (Note that $\{X_n\}_{n \ge 0}$ is a simple random walk started from 1 at time
   zero.)
   Let $T = \inf\{n : X_n = 0\}$. Show that T is a stopping time and that if $Y_n = X_{T \wedge n}$, then
   $\{Y_n\}_{n \ge 0}$ is a non-negative martingale and therefore, by Theorem 2.4.5, converges to
   a limit, Y_∞ as $n \to \infty$.
   Show that $\mathbb{E}[Y_n] = 1$ for all n, but that $Y_\infty = 0$. Why does this not contradict the
   conclusion of the Optional Stopping Theorem?

16   Recall Jensen's inequality: if g is a convex function and X a real-valued random
     variable then
$$\mathbb{E}\left[g(X)\right] \geq g\left(\mathbb{E}\left[X\right]\right).$$

Combine this with the characterisation (Example 2.4.7) of the discounted price
of an American call option on non-dividend-paying stock as the smallest $\mathbb{Q}$-
supermartingale that dominates $\{e^{-rn\delta t}(S_n - K)_+\}_{n \geq 0}$ to prove that the price of an
American call on non-dividend-paying stock is the same as that of a European call
with the same strike and maturity.

3 Brownian motion

Summary

Our discrete models are only a crude approximation to the way in which stock markets actually move. A better model would be one in which stock prices can change at any instant. As early as 1900 Bachelier, in his thesis 'La théorie de la spéculation', proposed Brownian motion as a model of the fluctuations of stock prices. Even today it is the building block from which we construct the basic reference model for a continuous time market. Before we can proceed further we must leave finance to define and construct Brownian motion.

Our first approach will be to continue the heuristic of §2.6 by considering Brownian motion as an 'infinitesimal' random walk in which smaller and smaller steps are taken at ever more frequent time intervals. This will lead us to a natural definition of the process. A formal construction, due to Lévy, will be given in §3.2, but this can safely be omitted. Next, §3.3 establishes some facts about the process that we shall require in later chapters. This material too can be skipped over and referred back to when it is used.

Just as discrete parameter martingales play a key rôle in the study of random walks, so for Brownian motion we shall use continuous time martingale theory to simplify a number of calculations; §3.4 extends our definitions and basic results on discrete parameter martingales to the continuous time setting.

3.1 Definition of the process

The easiest way to think about Brownian motion is as an 'infinitesimal random walk' and that is often how it arises in applications, so to motivate the formal definition we first study simple random walks.

A characterisation of simple random walks

We declared in Example 2.3.7 that the stochastic process $\{S_n\}_{n\geq 0}$ is a simple random walk under the measure $\mathbb{P}$ if $S_n = \sum_{i=1}^{n} \xi_i$ where the ξ_i can take only the values $\{-1, +1\}$ and are independent and identically distributed under $\mathbb{P}$. We concentrate on the *symmetric* case when

$$\mathbb{P}[\xi_i = -1] = \frac{1}{2} = \mathbb{P}[\xi_i = +1].$$

This process is often motivated as a model of the gains from repeated plays of a fair game. For example, suppose I play a game with a friend in which each play is equivalent to flipping a fair coin. If it comes up heads I pay her a dollar, otherwise she pays me a dollar. For each n, S_n models my net gain after n plays.

Recall from Exercise 13 of Chapter 2 that $\mathbb{E}[S_n] = 0$ and $var(S_n) = n$.

Lemma 3.1.1 $\{S_n\}_{n \geq 0}$ *is a* $\mathbb{P}$-*martingale (with respect to the natural filtration) and*

$$cov(S_n, S_m) = n \wedge m.$$

Proof: We checked in Example 2.3.7 that $\{S_n\}_{n \geq 0}$ is a $\mathbb{P}$-martingale. It remains to calculate the covariance.

$$
\begin{aligned}
cov(S_n, S_m) &= \mathbb{E}[S_n S_m] - \mathbb{E}[S_n]\mathbb{E}[S_m] \\
&= \mathbb{E}\left[\mathbb{E}[S_n S_m | \mathcal{F}_{m \wedge n}]\right] \qquad \text{(tower property)} \\
&= \mathbb{E}\left[S_{m \wedge n} \mathbb{E}[S_{m \vee n} | \mathcal{F}_{m \wedge n}]\right] \\
&= \mathbb{E}[S_{m \wedge n}^2] \qquad \text{(martingale property)} \\
&= var(S_{m \wedge n}) = m \wedge n.
\end{aligned}
$$

$\square$

As a result of the independence of the random variables $\{\xi_i\}_{i \geq 1}$, if $0 \leq i \leq j \leq k \leq l$, then $S_j - S_i$ is independent of $S_l - S_k$. More generally, if $0 \leq i_1 \leq i_2 \leq \cdots \leq i_n$, then $\{S_{i_r} - S_{i_{r-1}} : 1 \leq r \leq n\}$ are independent. Moreover, if $j - i = l - k = m$, say, then $S_j - S_i$ and $S_l - S_k$ both have the same distribution as S_m.

Notation: For two random variables X and Y we write

$$X \overset{\mathcal{D}}{=} Y$$

to mean that X and Y have the same distribution.
We also write $X \sim N(\mu, \sigma^2)$ to mean that X is normally distributed with mean μ and variance σ^2.

Combining the observations above we have

Lemma 3.1.2   *Under the measure* $\mathbb{P}$ *the process* $\{S_n\}_{n \geq 0}$ *has stationary, independent increments.*

Lemmas 3.1.1 and 3.1.2 are actually enough to *characterise* symmetric simple random walks.

Rescaling random walks

Recall that we want to think of Brownian motion as an infinitesimal random walk. In terms of our gambling game, the time interval between plays is δt and the stake is δx say, and we are thinking of both of these as 'tending to zero'. In order to obtain a non-trivial limit, there has to be a relationship between δt and δx. To see what this must be, we use the Central Limit Theorem (stated in §2.6). In our setting, $\mu = \mathbb{E}[\xi_i] = 0$ and $\sigma^2 = var(\xi_i) = 1$. Thus, taking $\delta t = 1/n$ and $\delta x = 1/\sqrt{n}$,

$$\mathbb{P}\left[\frac{S_n}{\sqrt{n}} \leq x\right] \rightarrow \int_{-\infty}^{x} \frac{1}{\sqrt{2\pi}} e^{-y^2/2} dy \text{ as } n \rightarrow \infty.$$

More generally,

$$\mathbb{P}\left[\frac{S_{[nt]}}{\sqrt{n}} \leq x\right] \rightarrow \int_{-\infty}^{x} \frac{1}{\sqrt{2\pi t}} e^{-y^2/2t} dy \text{ as } n \rightarrow \infty,$$

where $[nt]$ denotes the integer part of nt (Exercise 1). For the limiting process, at time t our net gain since time zero will be normally distributed with mean zero and variance t.

Definition of Brownian motion

Just as in our definition of a discrete time stochastic process, to define a continuous time stochastic process $\{X_t\}_{t\geq 0}$ (formally) requires a probability triple $(\Omega, \mathcal{F}, \mathbb{P})$ such that X_t is $\mathcal{F}$-measurable for all t. However, as in the discrete case, we shall rarely specify Ω explicitly.

Heuristically, passage to the limit in the random walk suggests that the following is a reasonable definition of Brownian motion.

Definition 3.1.3 (Brownian motion) *A real-valued stochastic process $\{W_t\}_{t\geq 0}$ is a $\mathbb{P}$-Brownian motion (or a $\mathbb{P}$-Wiener process) if for some real constant σ, under $\mathbb{P}$,*

1 *for each $s \geq 0$ and $t > 0$ the random variable $W_{t+s} - W_s$ has the normal distribution with mean zero and variance $\sigma^2 t$,*
2 *for each $n \geq 1$ and any times $0 \leq t_0 \leq t_1 \leq \cdots \leq t_n$, the random variables $\{W_{t_r} - W_{t_{r-1}}\}$ are independent,*
3 $W_0 = 0$,
4 W_t *is continuous in $t \geq 0$.*

Remarks: Conditions 1 and 2 ensure that, like its discrete counterpart, Brownian motion has stationary independent increments.

Condition 3 is a convention. Brownian motion started from x can be obtained as $\{x + W_t\}_{t\geq 0}$.

In a certain sense condition 4 is a consequence of the first three, but we should like to insist once and for all that all paths that our Brownian motion can follow are continuous.   □

The parameter σ^2 is known as the *variance* parameter. By scaling of the normal distribution it is immediate that $\{W_{t/\sigma}\}_{t\geq 0}$ is a Brownian motion with variance parameter one.

Definition 3.1.4   *The process with* $\sigma^2 = 1$ *is called* standard *Brownian motion.*

Assumption: Unless otherwise stated we shall always assume that $\sigma^2 = 1$.

Combining conditions 1 and 2 of Definition 3.1.3, we can write down the *transition probabilities* of standard Brownian motion.

$$\mathbb{P}\big[W_{t_n} \leq x_n | W_{t_i} = x_i, 0 \leq i \leq n-1\big] = \mathbb{P}\big[W_{t_n} - W_{t_{n-1}} \leq x_n - x_{n-1}\big]$$
$$= \int_{-\infty}^{x_n - x_{n-1}} \frac{1}{\sqrt{2\pi(t_n - t_{n-1})}} \exp\left(-\frac{u^2}{2(t_n - t_{n-1})}\right) du.$$

Notation: We write $p(t, x, y)$ for the *transition density*

$$p(t, x, y) = \frac{1}{\sqrt{2\pi t}} \exp\left(-\frac{(x-y)^2}{2t}\right).$$

This is the probability density function of the random variable W_{t+s} conditional on $W_s = x$.

For $0 = t_0 \leq t_1 \leq t_2 \leq \cdots \leq t_n$, writing $x_0 = 0$, the joint probability density function of $W_{t_1}, \ldots, W_{t_n}$ can also be written down explicitly as

$$f(x_1, \ldots, x_n) = \prod_{1}^{n} p(t_j - t_{j-1}, x_{j-1}, x_j).$$

The joint distributions of $W_{t_1}, \ldots, W_{t_n}$ for each $n \geq 1$ and all $t_1, \ldots, t_n$ are called the *finite dimensional distributions* of the process.

The following analogue of Lemma 3.1.1 is immediate.

Lemma 3.1.5   *For any* $s, t > 0$,

1   $\mathbb{E}\big[W_{t+s} - W_s | \{W_r\}_{0 \leq r \leq s}\big] = 0$,
2   $cov(W_s, W_t) = s \wedge t$.

In fact since the multivariate normal distribution is determined by its means and covariances and normally distributed random variables are independent if and only

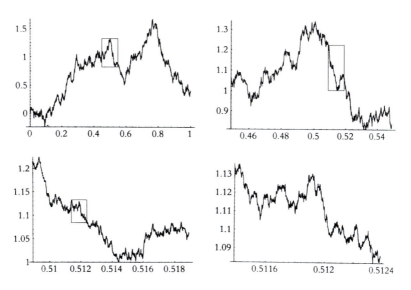

Figure 3.1   Zooming in on Brownian motion.

if their covariances are zero, this, combined with continuity of paths, characterises standard Brownian motion.

Behaviour of   Just because the sample paths of Brownian motion are continuous, it does not mean
Brownian   that they are nice in any other sense. In fact the behaviour of Brownian motion is
motion   distinctly odd. Here are just a few of its strange behavioural traits.

1   Although $\{W_t\}_{t \geq 0}$ is continuous everywhere, it is (with probability one) differentiable nowhere.

2   Brownian motion will eventually hit any and every real value no matter how large, or how negative. No matter how far above the axis, it will (with probability one) be back down to zero at some later time.

3   Once Brownian motion hits a value, it immediately hits it again *infinitely* often (and will continue to return after arbitrarily large times).

4   It doesn't matter what scale you examine Brownian motion on, it looks just the same. Brownian motion is a fractal.

Exercise 9 shows that the process cannot be differentiable at $t = 0$. We shall discuss some properties related to the hitting probabilities in §3.3 and in Exercise 8. The scaling alluded to in our last comment is formally proved in Proposition 3.3.7. It is really a consequence of the construction of the process. Figure 3.1 illustrates the result for a particular realisation of a Brownian path.

That such a bizarre process actually *exists* is far from obvious and so it is to this that we turn our attention in the next section.

3.2 Lévy's construction of Brownian motion

We have hinted that Brownian motion can be obtained as a limit of random walks. However, rather than chasing the technical details of the random walk construction, in this section we present an alternative construction due to Lévy. This can be omitted by readers willing to take existence of the process on trust.

A polygonal approximation

The idea is that we can simply produce a path of Brownian motion by direct polygonal interpolation. We require just one calculation.

Lemma 3.2.1   *Suppose that $\{W_t\}_{t \geq 0}$ is standard Brownian motion. Conditional on $W_{t_1} = x_1$, the probability density function of $W_{t_1/2}$ is*

$$p_{t_1/2}(x) \triangleq \sqrt{\frac{2}{\pi t_1}} \exp\left(-\frac{1}{2}\left(\frac{(x - \frac{1}{2}x_1)^2}{t_1/4}\right)\right).$$

In other words, the conditional distribution is a normally distributed random variable with mean $x_1/2$ and variance $t_1/4$. The proof is Exercise 11.

The construction: Without loss of generality we take the range of t to be $[0, 1]$. Lévy's construction builds (inductively) a polygonal approximation to the Brownian motion from a countable collection of *independent* normally distributed random variables with mean zero and variance one. We index them by the dyadic points of $[0, 1]$, a generic variable being denoted by $\xi\left(k2^{-n}\right)$ where $n \in \mathbb{N}$ and $k \in \{0, 1, \dots, 2^n\}$.

The induction begins with

$$X_1(t) = t\xi(1).$$

Thus X_1 is a linear function on $[0, 1]$.

The nth process, X_n, is linear in each interval $[(k-1)2^{-n}, k2^{-n}]$, is continuous in t and satisfies $X_n(0) = 0$. It is thus determined by the values $\{X_n(k2^{-n}), k = 1, \dots, 2^n\}$.

The inductive step: We take

$$X_{n+1}\left(2k2^{-(n+1)}\right) = X_n\left(2k2^{-(n+1)}\right) = X_n\left(k2^{-n}\right).$$

We now determine the appropriate value for $X_{n+1}\left((2k-1)2^{-(n+1)}\right)$. Conditional on $X_{n+1}\left(2k2^{-(n+1)}\right) - X_{n+1}\left(2(k-1)2^{-(n+1)}\right)$, Lemma 3.2.1 tells us that

$$X_{n+1}\left((2k-1)2^{-(n+1)}\right) - X_{n+1}\left(2(k-1)2^{-(n+1)}\right)$$

should be normally distributed with mean

$$\frac{1}{2}\left(X_{n+1}\left(2k2^{-(n+1)}\right) - X_{n+1}\left(2(k-1)2^{-(n+1)}\right)\right)$$

and variance $2^{-(n+2)}$.

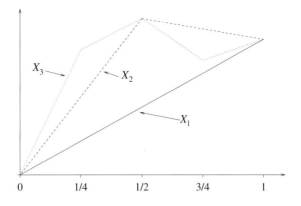

Figure 3.2　　Lévy's sequence of polygonal approximations to Brownian motion.

Now if $X \sim N(0, 1)$, then $aX + b \sim N(b, a^2)$ and so we take

$$X_{n+1}\left((2k-1)2^{-(n+1)}\right) - X_{n+1}\left(2(k-1)2^{-(n+1)}\right)$$
$$= 2^{-(n/2+1)}\xi\left((2k-1)2^{-(n+1)}\right)$$
$$+ \frac{1}{2}\left(X_{n+1}\left(2k2^{-(n+1)}\right) - X_{n+1}\left(2(k-1)2^{-(n+1)}\right)\right).$$

In other words

$$
\begin{aligned}
X_{n+1}\left((2k-1)2^{-(n+1)}\right) &= \frac{1}{2}X_n\left((k-1)2^{-n}\right) \\
&\quad + \frac{1}{2}X_n\left(k2^{-n}\right) + 2^{-(n/2+1)}\xi\left((2k-1)2^{-(n+1)}\right) \\
&= X_n\left((2k-1)2^{-(n+1)}\right) \\
&\quad + 2^{-(n/2+1)}\xi\left((2k-1)2^{-(n+1)}\right), \quad\quad\quad (3.1)
\end{aligned}
$$

where the last equality follows by linearity of X_n on $[(k-1)2^{-n}, k2^{-n}]$.
The construction is illustrated in Figure 3.2.

Convergence　Brownian motion will be the process constructed by letting n increase to infinity. To
to Brownian　check that it exists we need some technical lemmas. The proofs are adapted from
motion　Knight (1981).

Lemma 3.2.2

$$\mathbb{P}\left[\lim_{n\to\infty} X_n(t) \text{ exists for } 0 \le t \le 1 \text{ uniformly in } t\right] = 1.$$

Proof:　Notice that $\max_t |X_{n+1}(t) - X_n(t)|$ will be attained at a vertex, that is for

$t \in \{(2k-1)2^{-(n+1)}: k = 1, 2, \dots, 2^n\}$ and using (3.1)

$$\mathbb{P}\left[\max_t |X_{n+1}(t) - X_n(t)| \geq 2^{-n/4}\right]$$

$$= \mathbb{P}\left[\max_{1 \leq k \leq 2^n} \xi\left((2k-1)2^{-(n+1)}\right) \geq 2^{n/4+1}\right]$$

$$\leq 2^n \mathbb{P}\left[\xi(1) \geq 2^{n/4+1}\right].$$

Now using the result of Exercise 7 (with $t = 1$), for $x > 0$

$$\mathbb{P}\left[\xi(1) \geq x\right] \leq \frac{1}{x\sqrt{2\pi}} e^{-x^2/2},$$

and combining this with the fact that

$$\exp\left(-2^{(n/2+1)}\right) < 2^{-2n+2},$$

we obtain that for $n \geq 4$

$$2^n \mathbb{P}\left[\xi(1) \geq 2^{n/4+1}\right] \leq \frac{2^n}{2^{n/4+1}} \frac{1}{\sqrt{2\pi}} \exp\left(-2^{(n/2+1)}\right) \leq \frac{2^n}{2^{n/4+1}} 2^{-2n+2} < 2^{-n}.$$

Consider now for $k > n \geq 4$

$$\mathbb{P}\left[\max_t |X_k(t) - X_n(t)| \geq 2^{-n/4+3}\right] = 1 - \mathbb{P}\left[\max_t |X_k(t) - X_n(t)| \leq 2^{-n/4+3}\right]$$

and

$$\mathbb{P}\left[\max_t |X_k(t) - X_n(t)| \leq 2^{-n/4+3}\right]$$

$$\geq \mathbb{P}\left[\sum_{j=n}^{k-1} \max_t |X_{j+1}(t) - X_j(t)| \leq 2^{-n/4+3}\right]$$

$$\geq \mathbb{P}\left[\max_t |X_{j+1}(t) - X_j(t)| \leq 2^{-j/4}, j = n, \dots, k-1\right]$$

$$\geq 1 - \sum_{j=n}^{k-1} 2^{-j} \geq 1 - 2^{-n+1}.$$

Finally we have that

$$\mathbb{P}\left[\max_t |X_k(t) - X_n(t)| \geq 2^{-n/4+3}\right] \leq 2^{-n+1},$$

for all $k \geq n$. The events on the left are increasing (since the maximum can only increase by the addition of a new vertex) so

$$\mathbb{P}\left[\max_t |X_k(t) - X_n(t)| \geq 2^{-n/4+3} \text{ for some } k > n\right] \leq 2^{-n+1}.$$

In particular, for $\epsilon > 0$,

$$\lim_{n \to \infty} \mathbb{P}\left[\text{For some } k > n \text{ and } t \leq 1, |X_k(t) - X_n(t)| \geq \epsilon\right] = 0,$$

which proves the lemma.   □

To complete the proof of existence of the Brownian motion, we must check the following.

Lemma 3.2.3   Let $X(t) = \lim_{n \to \infty} X_n(t)$ if the limit exists uniformly and 0 otherwise. Then $X(t)$ satisfies the conditions of Definition 3.1.3 (for t restricted to $[0, 1]$).

Proof: By construction, the properties 1–3 of Definition 3.1.3 hold for the approximation $X_n(t)$ restricted to $T_n = \{k2^{-n} : k = 0, 1, \dots, 2^n\}$. Since we don't change X_k on T_n for $k > n$, the same must be true for X on $\bigcup_{n=1}^{\infty} T_n$. A uniform limit of continuous functions is continuous, so condition 4 holds and now by approximation of any $0 \le t_1 \le t_2 \le \cdots \le t_n \le 1$ from within the dense set $\bigcup_{n=1}^{\infty} T_n$ we see that in fact all four properties hold without restriction for $t \in [0, 1]$. $\square$

3.3   The reflection principle and scaling

Having proved that Brownian motion actually exists, we now turn to some calculations. These will amount to no more than a small bag of tricks for us to call upon in later chapters. There are many texts devoted exclusively to Brownian motion where the reader can gain a more extensive repertoire.

Stopping times

By its very construction, Brownian motion has no memory. That is, if $\{W_t\}_{t \ge 0}$ is a Brownian motion and $s \ge 0$ is any fixed time, then $\{W_{t+s} - W_s\}_{t \ge 0}$ is also a Brownian motion, independent of $\{W_r\}_{0 \le r \le s}$. What is also true is that for certain *random* times, T, the process $\{W_{T+t} - W_T\}_{t \ge 0}$ is again a standard Brownian motion and is independent of $\{W_s : 0 \le s \le T\}$. We have already encountered such random times in the context of discrete parameter martingales.

Definition 3.3.1   *A stopping time T for the process $\{W_t\}_{t \ge 0}$ is a random time such that for each t, the event $\{T \le t\}$ depends only on the history of the process up to and including time t.*

In other words, by observing the Brownian motion up until time t, we can determine whether or not $T \le t$.

We shall encounter stopping times only in the context of *hitting times*. For fixed a, the hitting time of level a is defined by

$$T_a = \inf\{t \ge 0 : W_t = a\}.$$

We take $T_a = \infty$ if a is never reached. It is easy to see that T_a is a stopping time since, by continuity of the paths,

$$\{T_a \le t\} = \{W_s = a \text{ for some } s, 0 \le s \le t\},$$

which depends only on $\{W_s, 0 \le s \le t\}$. Notice that, again by continuity, if $T_a < \infty$, then $W_{T_a} = a$.

Just as for random walks, an example of a random time that is *not* a stopping time is the *last* time that the process hits some level.

The reflection principle

Not surprisingly, there is often much to be gained from exploiting the symmetry inherent in Brownian motion. As a warm-up we calculate the distribution of T_a.

Lemma 3.3.2   *Let $\{W_t\}_{t\geq 0}$ be a $\mathbb{P}$-Brownian motion started from $W_0 = 0$ and let $a > 0$; then*

$$\mathbb{P}[T_a < t] = 2\mathbb{P}[W_t > a].$$

Proof: If $W_t > a$, then by continuity of the Brownian path, $T_a < t$. Moreover, since T_a is a stopping time, $\{W_{t+T_a} - W_{T_a}\}_{t\geq 0}$ is a Brownian motion, so, by symmetry, $\mathbb{P}[W_t - W_{T_a} > 0|T_a < t] = 1/2$. Thus

$$
\begin{aligned}
\mathbb{P}[W_t > a] &= \mathbb{P}[T_a < t, W_t - W_{T_a} > 0] \\
&= \mathbb{P}[T_a < t]\mathbb{P}[W_t - W_{T_a} > 0|T_a < t] \\
&= \frac{1}{2}\mathbb{P}[T_a < t].
\end{aligned}
$$

□

A more refined version of this idea is the following.

Lemma 3.3.3 (The reflection principle)   *Let $\{W_t\}_{t\geq 0}$ be a standard Brownian motion and let T be a stopping time. Define*

$$\tilde{W}_t = \begin{cases} W_t, & t \leq T, \\ 2W_T - W_t, & t > T; \end{cases}$$

then $\{\tilde{W}_t\}_{t\geq 0}$ is also a standard Brownian motion.

Notice that if $T = T_a$, then the operation $W_t \mapsto \tilde{W}_t$ amounts to reflecting the portion of the path after the first hitting time on a in the line $x = a$ (see Figure 3.3). We don't prove the general form of the reflection principle here. Instead we put it into action. The following result will be the key to pricing certain *barrier options* in Chapter 6.

Lemma 3.3.4 (Joint distribution of Brownian motion and its maximum)   *Let $M_t = \max_{0\leq s\leq t} W_s$, the maximum level reached by Brownian motion in the time interval $[0, t]$. Then for $a > 0$, $a \geq x$ and all $t \geq 0$,*

$$\mathbb{P}[M_t \geq a, W_t \leq x] = 1 - \Phi\left(\frac{2a - x}{\sqrt{t}}\right),$$

where

$$\Phi(x) = \int_{-\infty}^{x} \frac{1}{\sqrt{2\pi}} e^{-u^2/2} du$$

is the standard normal distribution function.

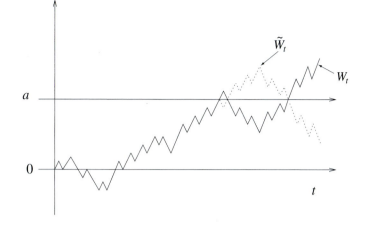

Figure 3.3   The reflection principle when $T = T_a$.

Proof: Notice that $M_t \geq 0$ and is non-decreasing in t and if, for $a > 0$, T_a is defined to be the first hitting time of level a, then $\{M_t \geq a\} = \{T_a \leq t\}$. Taking $T = T_a$ in the reflection principle, for $a \geq 0$, $a \geq x$ and $t \geq 0$,

$$
\begin{aligned}
\mathbb{P}[M_t \geq a, W_t \leq x] &= \mathbb{P}[T_a \leq t, W_t \leq x] \\
&= \mathbb{P}[T_a \leq t, 2a - x \leq \tilde{W}_t] \\
&= \mathbb{P}[2a - x \leq \tilde{W}_t] \\
&= 1 - \Phi\left(\frac{2a - x}{\sqrt{t}}\right).
\end{aligned}
$$

In the third equality we have used the fact that if $\tilde{W}_t \geq 2a - x$ then necessarily $\{\tilde{W}_s\}_{s \geq 0}$, and consequently $\{W_s\}_{s \geq 0}$, has hit level a *before* time t. □

Hitting a sloping line   For pricing a perpetual American put option in Chapter 6 we shall use the following result.

Proposition 3.3.5   *Set $T_{a,b} = \inf\{t \geq 0 : W_t = a + bt\}$, where $T_{a,b}$ is taken to be infinite if no such time exists. Then for $\theta > 0$, $a > 0$ and $b \geq 0$*

$$
\mathbb{E}\left[\exp\left(-\theta T_{a,b}\right)\right] = \exp\left(-a\left(b + \sqrt{b^2 + 2\theta}\right)\right).
$$

Proof: We defer the proof of the special case $b = 0$ until Proposition 3.4.9 when we shall have powerful martingale machinery to call upon. Here, assuming that result, we deduce the general result.

Fix $\theta > 0$, and for $a > 0$, $b \geq 0$, set

$$
\psi(a, b) = \mathbb{E}\left[e^{-\theta T_{a,b}}\right].
$$

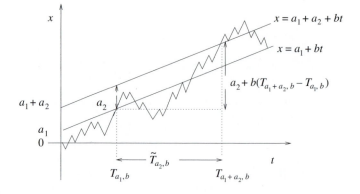

Figure 3.4   In the notation of Proposition 3.3.5, $T_{a_1+a_2,b} = T_{a_1,b} + \tilde{T}_{a_2,b}$ where $\tilde{T}_{a_2,b}$ has the same distribution as $T_{a_2,b}$.

Now take any two values for a, a_1 and a_2 say, and notice (see Figure 3.4) that

$$T_{a_1+a_2,b} = T_{a_1,b} + \left(T_{a_1+a_2,b} - T_{a_1,b}\right) \overset{D}{=} T_{a_1,b} + \tilde{T}_{a_2,b},$$

where $\tilde{T}_{a_2,b}$ is independent of $T_{a_1,b}$ and has the same distribution as $T_{a_2,b}$. In other words,

$$\psi(a_1 + a_2, b) = \psi(a_1, b)\psi(a_2, b),$$

and this implies that

$$\psi(a, b) = e^{-k(b)a},$$

for some function $k(b)$.

Since $b \geq 0$, the process must hit level a *before* it can hit the line $a + bt$. We use this to break $T_{a,b}$ into two parts; see Figure 3.5. Writing f_{T_a} for the probability density function of the random variable T_a and conditioning on T_a, we obtain

$$
\begin{aligned}
\psi(a, b) &= \int_0^\infty f_{T_a}(t) \mathbb{E}\left[e^{-\theta T_{a,b}}\,\Big|\, T_a = t\right] dt \\
&= \int_0^\infty f_{T_a}(t) e^{-\theta t} \mathbb{E}\left[e^{-\theta T_{bt,b}}\right] dt \\
&= \int_0^\infty f_{T_a}(t) e^{-\theta t} e^{-k(b)bt}\, dt \\
&= \mathbb{E}\left[e^{-(\theta + k(b)b)T_a}\right] \\
&= \exp\left(-a\sqrt{2(\theta + k(b)b)}\right).
\end{aligned}
$$

We now have two expressions for $\psi(a, b)$. Equating them gives

$$k^2(b) = 2\theta + 2k(b)b.$$

Since for $\theta > 0$ we must have $\psi(a, b) \leq 1$, we choose

$$k(b) = b + \sqrt{b^2 + 2\theta},$$

which completes the proof.   □

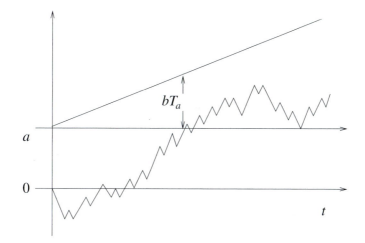

Figure 3.5   In the notation of Proposition 3.3.5, $T_{a,b} = T_a + \tilde{T}_{bT_a,b}$ where $\tilde{T}_{bT_a,b}$ has the same distribution as $T_{bT_a,b}$.

Definition 3.3.6   *For a real constant μ, we refer to the process $W_t^\mu = W_t + \mu t$ as a Brownian motion with drift μ.*

In the notation above, $T_{a,b}$ is the first hitting time of the level a by a Brownian motion with drift $-b$.

Transformation and scaling of Brownian motion

We conclude this section with the following useful result.

Proposition 3.3.7
If $\{W_t\}_{t\geq 0}$ is a standard Brownian motion, then so are

1   $\{cW_{t/c^2}\}_{t\geq 0}$ *for any real c,*
2   $\{tW_{1/t}\}_{t\geq 0}$ *where $tW_{1/t}$ is taken to be zero when $t = 0$,*
3   $\{W_s - W_{s-t}\}_{0\leq t\leq s}$ *for any fixed $s \geq 0$.*

Proof: The proofs of 1–3 are similar. For example in the case of 2, it is clear that $tW_{1/t}$ has continuous sample paths (at least for $t > 0$) and that for any $t_1, \ldots, t_n$, the random variables $\{t_1 W_{1/t_1}, \ldots, t_n W_{1/t_n}\}$ have a multivariate normal distribution. We must just check that the covariance takes the right form, but

$$\mathbb{E}\left[sW_{1/s}tW_{1/t}\right] = st\mathbb{E}\left[W_{1/s}W_{1/t}\right] = st\left(\frac{1}{s} \wedge \frac{1}{t}\right) = s \wedge t,$$

and the proof is complete.   □

3.4   Martingales in continuous time

Just as in discrete time, the notion of a martingale plays a key rôle in our continuous time models.

Recall that in discrete time, a sequence $X_0, X_1, \ldots, X_n$ for which $\mathbb{E}[|X_r|] < \infty$ for each r is a martingale with respect to the filtration $\{\mathcal{F}_n\}_{n \geq 0}$ and a probability measure $\mathbb{P}$ if

$$\mathbb{E}\left[X_r | \mathcal{F}_{r-1}\right] = X_{r-1} \qquad \text{for all } r \geq 1.$$

We can make entirely analogous definitions in continuous time.

Filtrations

Definition 3.4.1 *Let $\mathcal{F}$ be a σ-field. We call $\{\mathcal{F}_t\}_{t \geq 0}$ a filtration if*

1 *$\mathcal{F}_t$ is a sub-σ-field of $\mathcal{F}$ for all t, and*
2 *$\mathcal{F}_s \subseteq \mathcal{F}_t$ for $s < t$.*

As in the discrete setting we are primarily concerned with the *natural* filtration, $\{\mathcal{F}_t^X\}_{t \geq 0}$, associated with a stochastic process $\{X_t\}_{t \geq 0}$. As before, $\mathcal{F}_t^X$ encodes the information generated by the stochastic process X on the interval $[0, t]$. That is $A \in \mathcal{F}_t^X$ if, based upon observations of the trajectory $\{X_s\}_{0 \leq s \leq t}$, it is possible to decide whether or not A has occurred.

> **Notation:** If the value of a stochastic variable Z can be completely determined given observations of the trajectory $\{X_s\}_{0 \leq s \leq t}$ then we write
>
> $$Z \in \mathcal{F}_t^X.$$

More than one process can be measurable with respect to the same filtration.

Definition 3.4.2 *If $\{Y_t\}_{t \geq 0}$ is a stochastic process such that we have $Y_t \in \mathcal{F}_t^X$ for all $t \geq 0$, then we say that $\{Y_t\}_{t \geq 0}$ is* adapted *to the filtration $\{\mathcal{F}_t^X\}_{t \geq 0}$.*

Example 3.4.3

1 *The stochastic process*
$$Z_t = \int_0^t X_s ds$$
is adapted to $\{\mathcal{F}_t^X\}_{t \geq 0}$.
2 *The process $M_t = \max_{0 \leq s \leq t} W_s$ is adapted to the filtration $\{\mathcal{F}_t^W\}_{t \geq 0}$.*
3 *The stochastic process $Z_t \triangleq W_{t+1}^2 - W_t^2$ is* not *adapted to the filtration generated by $\{W_t\}_{t \geq 0}$.*

Notice that just as in the discrete world we have divorced the rôles of the stochastic process and the probability measure. Thus a process may be a Brownian motion under the probability measure $\mathbb{P}$, but the *same* process not be a Brownian motion under a different measure $\mathbb{Q}$.

Martingales

Definition 3.4.4 *Let $(\Omega, \mathcal{F}, \mathbb{P})$ be a probability space with filtration $\{\mathcal{F}_t\}_{t \geq 0}$. A family $\{M_t\}_{t \geq 0}$ of random variables on this space with $\mathbb{E}[|M_t|] < \infty$ for all $t \geq 0$ is* *a* $(\mathbb{P}, \{\mathcal{F}_t\}_{t\geq 0})$-martingale *if it is adapted to* $\{\mathcal{F}_t\}_{t\geq 0}$ *and for any* $s \leq t$,

$$\mathbb{E}^{\mathbb{P}}\left[M_t \mid \mathcal{F}_s\right] = M_s.$$

By restricting the conditions to $t \in [0, T]$, *we define martingales parametrised by* $[0, T]$.

Generally we shall be sloppy about specifying the filtration. In all of our examples there will be a Brownian motion around and it will be implicit that the filtration is that generated by the Brownian motion.

A more general notion is that of local martingale.

Definition 3.4.5   *A process* $\{X_t\}_{t\geq 0}$ *is a local* $(\mathbb{P}, \{\mathcal{F}_t\}_{t\geq 0})$-martingale *if there is a sequence of* $\{\mathcal{F}_t\}_{t\geq 0}$-*stopping times* $\{T_n\}_{n\geq 1}$ *such that* $\{X_{t\wedge T_n}\}_{t\geq 0}$ *is a* $(\mathbb{P}, \{\mathcal{F}_t\}_{t\geq 0})$-*martingale for each n and*

$$\mathbb{P}\left[\lim_{n\to\infty} T_n = \infty\right] = 1.$$

All martingales are local martingales but the converse is false. It is because of this distinction that we impose boundedness conditions in many of our results of Chapter 4.

Lemma 3.4.6   *Let* $\{W_t\}_{t\geq 0}$ *generate the filtration* $\{\mathcal{F}_t\}_{t\geq 0}$. *If* $\{W_t\}_{t\geq 0}$ *is a standard Brownian motion under the probability measure* $\mathbb{P}$, *then*

1   W_t *is a* $(\mathbb{P}, \{\mathcal{F}_t\}_{t\geq 0})$-*martingale,*
2   $W_t^2 - t$ *is a* $(\mathbb{P}, \{\mathcal{F}_t\}_{t\geq 0})$-*martingale,*
3

$$\exp\left(\sigma W_t - \frac{\sigma^2}{2}t\right)$$

is a $(\mathbb{P}, \{\mathcal{F}_t\}_{t\geq 0})$-*martingale, called an* exponential martingale.

Proof: The proofs are all rather similar. For example, consider $M_t = W_t^2 - t$. Evidently $\mathbb{E}[|M_t|] < \infty$. Now

$$
\begin{aligned}
\mathbb{E}\left[W_t^2 - W_s^2 \mid \mathcal{F}_s\right] &= \mathbb{E}\left[(W_t - W_s)^2 + 2W_s(W_t - W_s) \mid \mathcal{F}_s\right] \\
&= \mathbb{E}\left[(W_t - W_s)^2 \mid \mathcal{F}_s\right] + 2W_s\mathbb{E}\left[(W_t - W_s) \mid \mathcal{F}_s\right] \\
&= t - s.
\end{aligned}
$$

Thus

$$
\begin{aligned}
\mathbb{E}\left[W_t^2 - t \mid \mathcal{F}_s\right] &= \mathbb{E}\left[W_t^2 - W_s^2 + W_s^2 - (t-s) - s \mid \mathcal{F}_s\right] \\
&= (t-s) + W_s^2 - (t-s) - s = W_s^2 - s.
\end{aligned}
$$

□

Optional
stopping

What we should really like is the continuous time analogue of the Optional Stopping Theorem. In general, we have to be a little careful (see Exercise 17 for what can go wrong). Problems arise if the sample paths of our martingale are not sufficiently 'nice'. In all our examples the stochastic process will have càdlàg sample paths.

Definition 3.4.7   *The function $f : \mathbb{R} \to \mathbb{R}$ is càdlàg if it is right continuous with left limits.*

In particular, continuous functions are automatically càdlàg (continues à droite, limites à gauche).

Theorem 3.4.8 (Optional Stopping Theorem)   *If $\{M_t\}_{t \geq 0}$ is a càdlàg martingale with respect to the probability measure $\mathbb{P}$ and the filtration $\{\mathcal{F}_t\}_{t \geq 0}$ and if τ_1 and τ_2 are two stopping times such that $\tau_1 \leq \tau_2 \leq K$ where K is a finite real number, then*

$$\mathbb{E}\left[|M_{\tau_2}|\right] < \infty$$

and

$$\mathbb{E}\left[M_{\tau_2} \mid \mathcal{F}_{\tau_1}\right] = M_{\tau_1}, \qquad \mathbb{P}\text{-}a.s.$$

Remarks:

1   The term 'a.s.' (almost surely) means with ($\mathbb{P}$-) probability one.
2   Notice in particular that if τ is a bounded stopping time then $\mathbb{E}[M_\tau] = \mathbb{E}[M_0]$.

□

Brownian
hitting time
distribution

Just as in the discrete case the Optional Stopping Theorem will be a powerful tool. We illustrate by calculating the moment generating function for the hitting time T_a of level a by Brownian motion. (This result was essential to our proof of Proposition 3.3.5.)

Proposition 3.4.9   *Let $\{W_t\}_{t \geq 0}$ be a Brownian motion and let $T_a = \inf\{s \geq 0 : W_s = a\}$ (or infinity if that set is empty). Then for $\theta > 0$,*

$$\mathbb{E}\left[e^{-\theta T_a}\right] = e^{-\sqrt{2\theta}|a|}.$$

Proof: We assume that $a \geq 0$. (The case $a < 0$ follows by symmetry.) We should like to apply the Optional Stopping Theorem to the martingale

$$M_t = \exp\left(\sigma W_t - \frac{1}{2}\sigma^2 t\right)$$

and the random time T_a, but we encounter a familiar obstacle. We *cannot* apply the Theorem directly to T_a as it may not be bounded. Instead we take $\tau_1 = 0$ and $\tau_2 = T_a \wedge n$. This gives us that

$$\mathbb{E}\left[M_{T_a \wedge n}\right] = 1.$$

So

$$1 = \mathbb{E}\left[M_{T_a \wedge n}\right] = \mathbb{E}\left[M_{T_a \wedge n} \,\middle|\, T_a < n\right]\mathbb{P}\left[T_a < n\right]$$
$$+ \mathbb{E}\left[M_{T_a \wedge n} \,\middle|\, T_a > n\right]\mathbb{P}\left[T_a > n\right]. \qquad (3.2)$$

Now, by Lemma 3.3.2 and the result of Exercise 7,

$$\mathbb{P}\left[T_a < n\right] \to 1 \qquad \text{as } n \to \infty.$$

Also, if $T_a < \infty$, $\lim_{n \to \infty} M_{T_a \wedge n} = M_{T_a}$, whereas if $T_a = \infty$, $W_t \le a$ for all t and so $\lim_{n \to \infty} M_{T_a \wedge n} = 0$. Letting $n \to \infty$ in equation (3.2) then yields

$$\mathbb{E}\left[M_{T_a}\right] = 1.$$

Taking $\sigma^2 = 2\theta$ completes the proof. $\qquad\qquad\qquad\qquad\qquad\qquad \square$

Dominated Convergence Theorem

Arguments of this type are often simplified by an application of the Dominated Convergence Theorem.

Theorem 3.4.10 (Dominated Convergence Theorem)   *Let* $\{Z_n\}_{n \ge 1}$ *be a sequence of random variables with* $\lim_{n \to \infty} Z_n = Z$. *If there is a random variable* Y *with* $|Z_n| < Y$ *for all* n *and* $\mathbb{E}[Y] < \infty$, *then*

$$\mathbb{E}[Z] = \lim_{n \to \infty} \mathbb{E}[Z_n].$$

In the proof of Proposition 3.4.9, since

$$0 \le M_{T_a \wedge n} = \exp\left(\sigma W_{T_a \wedge n} - \frac{1}{2}\sigma^2 (T_a \wedge n)\right) \le \exp\left(\sigma a\right),$$

we could take the *constant* $e^{\sigma a}$ as the dominating random variable Y.

Exercises

1   Suppose that $\{S_n\}_{n \ge 0}$ is a symmetric simple random walk under $\mathbb{P}$. Show that

$$\mathbb{P}\left[\frac{S_{[nt]}}{\sqrt{n}} \le x\right] \to \int_{-\infty}^{x} \frac{1}{\sqrt{2\pi t}} \exp\left(-\frac{y^2}{2t}\right) dy$$

as $n \to \infty$ where $[nt]$ is the integer part of nt.

2   Let Z be normally distributed with mean zero and variance one under the measure $\mathbb{P}$. What is the distribution of $\sqrt{t}Z$? Is the process $X_t = \sqrt{t}Z$ a Brownian motion?

3   Suppose that W_t and $\tilde{W}_t$ are independent Brownian motions under the measure $\mathbb{P}$ and let $\rho \in [-1, 1]$ be a constant. Is the process $X_t = \rho W_t + \sqrt{1 - \rho^2}\tilde{W}_t$ a Brownian motion?

4   Let $\{W_t\}_{t\geq 0}$ be standard Brownian motion under the measure $\mathbb{P}$. Which of the following are $\mathbb{P}$-Brownian motions?

(a) $\{-W_t\}_{t\geq 0}$,
(b) $\{cW_{t/c^2}\}_{t\geq 0}$, where c is a constant,
(c) $\{\sqrt{t}W_1\}_{t\geq 0}$,
(d) $\{W_{2t} - W_t\}_{t\geq 0}$.

Justify your answers.

5   Suppose that X is normally distributed with mean μ and variance σ^2. Calculate

$$\mathbb{E}\left[e^{\theta X}\right]$$

and hence evaluate $\mathbb{E}\left[X^4\right]$.

6   Prove Lemma 3.4.6.3.

7   Prove that if $\{W_t\}_{t\geq 0}$ is standard Brownian motion under $\mathbb{P}$ then, for $x > 0$,

$$\mathbb{P}\left[W_t \geq x\right] \equiv \int_x^\infty \frac{1}{\sqrt{2\pi t}}e^{-y^2/2t}dy \leq \frac{\sqrt{t}}{x\sqrt{2\pi}}e^{-x^2/2t}.$$

[Hint: Integrate by parts.]

8   Let $\{W_t\}_{t\geq 0}$ be standard Brownian motion under $\mathbb{P}$. Let $Z = \sup_t W_t$. Evidently, for any $c > 0$, cZ has the same distribution as Z. Deduce that, with probability one, $Z \in \{0, \infty\}$. Let $p = \mathbb{P}[Z = 0]$. By conditioning on the event $\{W_1 \leq 0\}$, prove that

$$\mathbb{P}[Z = 0] \leq \mathbb{P}[W_1 \leq 0]\,\mathbb{P}[Z = 0],$$

and hence $p = 0$. Deduce that

$$\mathbb{P}\left[\sup_t W_t = +\infty, \inf_t W_t = -\infty\right] = 1.$$

9   Deduce from the result of Exercise 8 and the result of Proposition 3.3.7.2 that

$$\mathbb{P}\left[\text{For each } \epsilon > 0, \exists s, t \leq \epsilon \text{ such that } W_s < 0 < W_t\right] = 1.$$

Deduce that if $\{W_t\}_{t\geq 0}$ is differentiable at zero, then the derivative must be zero and hence $|W_t| \leq t$ for all sufficiently small t. By considering $\tilde{W}_s \triangleq sW_{1/s}$, arrive at a contradiction and deduce that Brownian motion is *not* differentiable at zero.

10 Brownian motion is not going to be adequate as a stock market model. First, it has constant mean, whereas the stock of a company usually grows at some rate, if only due to inflation. Moreover, it may be too 'noisy' (that is the variance of the increments may be bigger than those observed for the stock) or not noisy enough. We can scale to change the 'noisiness' and we can artificially introduce a drift, but this still won't be a good model. Here is one reason why. Suppose that $\{W_t\}_{t\geq 0}$ is standard Brownian motion under $\mathbb{P}$. Define a new process $\{S_t\}_{t\geq 0}$ by $S_t = \mu t + \sigma W_t$ where $\sigma > 0$ and $\mu \in \mathbb{R}$ are constants. Show that for all values of $\sigma > 0$, $\mu \in \mathbb{R}$ and $T > 0$ there is a positive probability that S_T is negative.

11   Suppose that $\{W_t\}_{t\geq0}$ is standard Brownian motion. Prove that conditional on $W_{t_1} = x_1$, the probability density function of $W_{t_1/2}$ is

$$\sqrt{\frac{2}{\pi t_1}} \exp\left(-\frac{1}{2}\left(\frac{(x - \frac{1}{2}x_1)^2}{t_1/4}\right)\right).$$

12   Let $\{W_t\}_{t\geq0}$ be standard Brownian motion under $\mathbb{P}$. Let T_a be the 'hitting time of level a', that is

$$T_a = \inf\{t \geq 0 : W_t = a\}.$$

Then we proved in Proposition 3.4.9 that

$$\mathbb{E}\left[\exp\left(-\theta T_a\right)\right] = \exp\left(-a\sqrt{2\theta}\right).$$

Use this result to calculate

(a) $\mathbb{E}[T_a]$,
(b) $\mathbb{P}[T_a < \infty]$.

13   Let $\{W_t\}_{t\geq0}$ denote standard Brownian motion under $\mathbb{P}$ and define $\{M_t\}_{t\geq0}$ by

$$M_t = \max_{0\leq s\leq t} W_s.$$

Suppose that $x \geq a$. Calculate

(a) $\mathbb{P}[M_t \geq a, W_t \geq x]$,
(b) $\mathbb{P}[M_t \geq a, W_t \leq x]$.

14   Let $\{W_t\}_{t\geq0}$ be standard Brownian motion under $\mathbb{P}$. Let $T_{a,b}$ denote the hitting time of the sloping line $a + bt$. That is,

$$T_{a,b} = \inf\{t \geq 0 : W_t = a + bt\}.$$

We proved in Proposition 3.3.5 that for $\theta > 0$, $a > 0$ and $b \geq 0$

$$\mathbb{E}\left[\exp\left(-\theta T_{a,b}\right)\right] = \exp\left(-a\left(b + \sqrt{b^2 + 2\theta}\right)\right).$$

The aim of this question is to calculate the distribution of $T_{a,b}$, without inverting the Laplace transform. In what follows, $\phi(x) = \Phi'(x)$ and

$$\Phi(x) = \int_{-\infty}^{x} \frac{1}{\sqrt{2\pi}} \exp\left(-\frac{y^2}{2}\right) dy.$$

(a) Find $\mathbb{P}[T_{a,b} < \infty]$.
(b) Using the fact that $s W_{1/s}$ has the same distribution as W_s, show that

$$\mathbb{P}\left[T_{a,b} \leq t\right] = \mathbb{P}\left[W_s \geq as + b \text{ for some } s \text{ with } 1/t \leq s < \infty\right].$$

(c) By conditioning on the value of $W_{1/t}$, use the previous part to show that

$$\mathbb{P}\left[T_{a,b} \leq t\right] = \int_{-\infty}^{b+a/t} \mathbb{P}\left[T_{b-x+a/t,a} < \infty\right] \phi\left(\sqrt{t}x\right) dx + 1 - \Phi\left(\frac{a+bt}{\sqrt{t}}\right).$$

(d) Substitute for the probability in the integral and deduce that

$$\mathbb{P}\left[T_{a,b} \leq t\right] = e^{-2ab} \Phi\left(\frac{bt-a}{\sqrt{t}}\right) + 1 - \Phi\left(\frac{a+bt}{\sqrt{t}}\right).$$

15 Let $\{W_t\}_{t\geq0}$ be standard Brownian motion under the measure $\mathbb{P}$ and let $\{\mathcal{F}_t\}_{t\geq0}$ denote its natural filtration. Which of the following are $(\mathbb{P}, \{\mathcal{F}_t\}_{t\geq0})$-martingales?

(a) $\exp\left(\sigma W_t\right)$,
(b) cW_{t/c^2}, where c is a constant,
(c) $tW_t - \int_0^t W_s ds$.

16 Let $\{\mathcal{F}_t\}_{0\leq t\leq T}$ denote the natural filtration associated to a standard $\mathbb{P}$-Brownian motion, $\{W_t\}_{0\leq t\leq T}$. The result of Lemma 3.4.6.3 can be rewritten as

$$\mathbb{E}\left[\exp\left(\sigma W_t - \frac{1}{2}\sigma^2 t\right); A\right] = \exp\left(\sigma W_s - \frac{1}{2}\sigma^2 s\right) \mathbf{1}_A, \quad \text{for all } A \in \mathcal{F}_s.$$

Use differentiation under the integral sign to provide another proof that $\{W_t^2 - t\}_{t\geq0}$ is a $(\mathbb{P}, \{\mathcal{F}_t\}_{t\geq0})$-martingale and show that the following are also $(\mathbb{P}, \{\mathcal{F}_t\}_{t\geq0})$-martingales:

(a) $W_t^3 - 3tW_t$,
(b) $W_t^4 - 6tW_t^2 + 3t^2$.

17 Let $(\Omega, \mathcal{F}, \mathbb{P})$ be a probability space. Suppose that the real random variable $T : \Omega \to \mathbb{R}$ is uniformly distributed on $[0, 1]$ under the measure $\mathbb{P}$. Define $\{X_t\}_{t\geq0}$ by

$$X_t(\omega) = \begin{cases} 1, & T(\omega) = t, \\ 0, & T(\omega) \neq t. \end{cases}$$

Check that $\{X_t\}_{t\geq0}$ is a $\mathbb{P}$-martingale with respect to its own filtration. [Hint: Conditional expectation is only unique to within a random variable that is almost surely zero.]
Show that T is a stopping time for which the Optional Stopping Theorem fails.

18 As before, let T_a, T_b denote the first hitting times of levels a and b respectively of a $\mathbb{P}$-Brownian motion, $\{W_t\}_{t\geq0}$, but now W_0 is not necessarily zero (see the remarks after Definition 3.1.3). Prove that if $a < x < b$ then

$$\mathbb{P}[T_a < T_b| W_0 = x] = \frac{(b-x)}{(b-a)}.$$

[Hint: Mimic the proof of the corresponding result for random walk, cf. Proposition 2.4.4.]

19 Using the notation of Exercise 18, let $T = T_a \wedge T_b$. Prove that if $a < 0 < b$ then

$$\mathbb{E}[T| W_0 = 0] = -ab.$$

4 Stochastic calculus

Summary

Brownian motion is clearly inadequate as a market model, not least because it would predict negative stock prices. However, by considering *functions* of Brownian motion we can produce a wide class of potential models. The basic model underlying the Black–Scholes pricing theory, geometric Brownian motion, arises precisely in this way. It will inherit from the Brownian motion very irregular paths. In §4.1 we shall see why a stock price model with rough paths is forced upon us by arbitrage arguments. This is not in itself sufficient to justify the geometric Brownian motion model. However in §4.7 we provide a further argument that suggests that it is at least a sensible starting point. A more detailed discussion of the shortcomings of the geometric Brownian motion model is deferred until Chapter 7.

In order to study models built in this way, we need to develop a calculus based on Brownian motion. The Itô stochastic calculus is the main topic of this chapter. In §4.2 we define the Itô stochastic integral and then in §4.3 we derive the corresponding chain rule of stochastic calculus and learn how to integrate by parts.

Just as in the discrete world, there will be two key ingredients to pricing and hedging in the Black–Scholes framework. First we need to be able to *change the probability measure* so that discounted asset prices are martingales. The tool for doing this is the Girsanov Theorem of §4.5. The construction of the hedging portfolio depends on the continuous analogue of the Binomial Representation Theorem, the Martingale Representation Theorem of §4.6.

Again as in the discrete world, the pricing formula will be in the form of the discounted expected value of a claim. Black and Scholes obtained this result via a completely different argument (see Exercise 5 of Chapter 5) in which the price is obtained as the solution of a partial differential equation. The connection with the probabilistic approach is via the Feynman–Kac stochastic representation formula of §4.8 which exposes the intricate connection between stochastic differential equations and certain second order parabolic (deterministic) partial differential equations.

Once again our coverage of this material is necessarily rather sketchy. Even so readers eager to get back to some finance may wish to skip the proofs in this

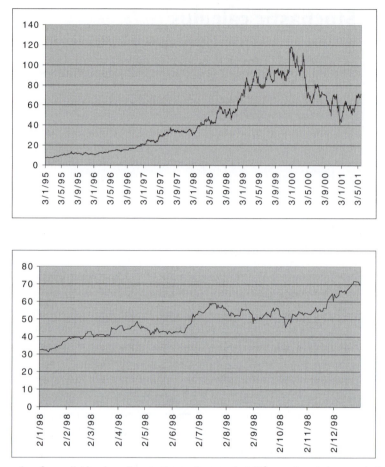

Figure 4.1   Two graphs of non-dividend-paying stock over long period ($6\frac{1}{3}$ years) and short period (1 year).

chapter. There is no shortage of excellent stochastic calculus texts to refer to. Some suggestions are included in the bibliography.

4.1   Stock prices are not differentiable

Figure 4.1 shows the Microsoft share price over $6\frac{1}{3}$ year and 1 year periods. It certainly doesn't look like a particularly nice function of time. Even over short time scales, the path followed by the price looks rough. There are many statistical studies that investigate the irregularity of paths of stock prices. In this section we explore through a purely mathematical argument of Lyons (1995) just how rough paths of our stock price model should be, at least under the assumption that we can trade continuously without incurring transaction costs and, as usual, that there are no arbitrage opportunities. We continue to suppose that our market contains a riskless cash bond.

Quantifying
roughness

First we need a means of quantifying 'roughness'. For a function $f : [0, T] \to \mathbb{R}$, its variation is defined in terms of partitions.

Definition 4.1.1   *Let π be a partition of $[0, T]$, $N(\pi)$ the number of intervals that make up π and $\delta(\pi)$ be the mesh of π (that is the length of the largest interval in the partition). Write $0 = t_0 < t_1 < \cdots < t_{N(\pi)} = T$ for the endpoints of the intervals of the partition. Then the variation of f is*

$$\lim_{\delta \to 0} \left\{ \sup_{\pi:\delta(\pi)=\delta} \sum_{j=1}^{N(\pi)} |f(t_j) - f(t_{j-1})| \right\}.$$

If the function is 'nice', for example differentiable, then it has bounded variation. Our 'rough' paths will have *unbounded* variation. To quantify roughness we can extend the idea of variation to that of p-variation.

Definition 4.1.2   *In the notation of Definition 4.1.1, the p-variation of a function $f : [0, T] \to \mathbb{R}$ is defined as*

$$\lim_{\delta \to 0} \left\{ \sup_{\pi:\delta(\pi)=\delta} \sum_{j=1}^{N(\pi)} |f(t_j) - f(t_{j-1})|^p \right\}.$$

Notice that if $p > 1$ the p-variation will be finite for functions that are much rougher than those for which the variation is bounded. For example, roughly speaking, finite 2-variation will follow if the fluctuation of the function over an interval of order δ is order $\sqrt{\delta}$.

Bounded
variation and
arbitrage

We now argue that if stock prices had bounded variation, then either they would be constant multiples of the riskless cash bond or (provided we can trade continuously and there are no transaction costs) there would be unbounded arbitrage opportunities.

In the discrete time world of Chapter 2 we showed (equation (2.5)) that if a portfolio consisting of ϕ_{i+1} units of stock and ψ_{i+1} cash bonds over the time interval $[i\delta t, (i + 1)\delta t)$ is self-financing, then its discounted value at time $N\delta t = T$ is

$$\tilde{V}_N = V_0 + \sum_{j=0}^{N-1} \phi_{j+1} \left(\tilde{S}_{j+1} - \tilde{S}_j \right). \tag{4.1}$$

Here ϕ_{j+1} is known at time $j\delta t$, but is typically a function of $\tilde{S}_j$. In our continuous world, we can let the trading interval δt tend to zero and, if the discounted stock price process has bounded variation, as $\delta t \downarrow 0$ the Riemann sum in (4.1) will converge to the Riemann integral

$$\int_0^T \phi_t \left(\tilde{S}_t \right) d\tilde{S}_t$$

where ϕ_t denotes our stock holding at time t. This says that for any choice of $\{\phi_t(\cdot)\}_{0 \le t \le T}$, we can construct a self-financing portfolio whose discounted value at time T is

$$V_0 + \int_0^T \phi_t(\tilde{S}_t) d\tilde{S}_t.$$

Now choose a differentiable function $F(x)$ that is small near $x = S_0$ and very large everywhere else. Then by investing $F(S_0)$ at time zero and holding a self-financing portfolio with $\phi(\tilde{S}_t)$ units of stock at time t, where $\phi(x) = F'(x)$, we generate a portfolio at time T whose discounted value is

$$F(S_0) + \int_0^T F'(\tilde{S}_t)d\tilde{S}_t,$$

which, by the Fundamental Theorem of Calculus, is $F(\tilde{S}_T)$.

We only have to wait for the discounted stock price to move away from S_0 to generate a lot of wealth. For example, the strategy that holds $\left(\tilde{S}_t - \tilde{S}_0\right)$ units of stock at time t generates

$$e^{rT}\int_0^T (\tilde{S}_t - \tilde{S}_0)d\tilde{S}_t = e^{rT}\left(\tilde{S}_T - \tilde{S}_0\right)^2$$

units of wealth at time T (where we have multiplied by e^{rT} to 'undo' the discounting).

In the absence of arbitrage then we do not expect the paths of our stock price to have bounded variation. In fact, as Lyons points out, arguments of L C Young extend this. Again assuming continuous trading and no transaction costs, if the paths of the stock price have finite p-variation for some $p < 2$, then there are arbitrarily large profits to be made.

4.2 Stochastic integration

The work of §4.1 suggests that we should be looking for models in which the stock price has infinite p-variation for $p < 2$. A large class of such models can be constructed using Brownian motion as a building block, but this will require a new calculus. The paths of Brownian motion are too rough for the familiar Newtonian calculus to help us and, indeed, if it did the Fundamental Theorem of Calculus would once again lead us to discard Brownian motion as a basis for our models.

A differential equation for the stock price

The processes used to model stock prices are usually functions of one or more Brownian motions. Here, for simplicity, we restrict ourselves to functions of just one Brownian motion. The first thing that we should like to do is to write down a differential equation for the way in which the stock price evolves.

Suppose that the stock price is of the form $S_t = f(t, W_t)$. Using Taylor's Theorem (and assuming that f at least is 'nice'),

$$f(t + \delta t, W_{t+\delta t}) - f(t, W_t) = \delta t \dot{f}(t, W_t) + O(\delta t^2) + (W_{t+\delta t} - W_t) f'(t, W_t)$$
$$+ \frac{1}{2!}(W_{t+\delta t} - W_t)^2 f''(t, W_t) + \cdots$$

where we have used the notation

$$\dot{f}(t, x) = \frac{\partial f}{\partial t}(t, x), \quad f'(t, x) = \frac{\partial f}{\partial x}(t, x) \text{ and } f''(t, x) = \frac{\partial^2 f}{\partial x^2}(t, x).$$

Now in our usual derivation of the chain rule, when $\{W_t\}_{t \geq 0}$ is replaced by a bounded variation function, the last term on the right hand side is order $O(\delta t^2)$. However, for Brownian motion, we know that $\mathbb{E}[(W_{t+\delta t} - W_t)^2]$ is δt. Consequently we cannot ignore the term involving the second derivative. Of course, now we have a problem, because we must interpret the term involving the *first* derivative. If $(W_{t+\delta t} - W_t)^2$ is $O(\delta t)$, then $(W_{t+\delta t} - W_t)$ should be $O(\sqrt{\delta t})$, which could lead to unbounded changes in $\{S_t\}_{t \geq 0}$ over a bounded time interval. However, things are not hopeless. The expected value of $W_{t+\delta t} - W_t$ is zero, and the fluctuations around zero are on the order of $\sqrt{\delta t}$. By comparison with the Central Limit Theorem, it is plausible that $S_t - S_0$ is a well-defined random variable. Assuming that we can make this rigorous, the differential equation governing $S_t = f(t, W_t)$ will take the form

$$dS_t = \dot{f}(t, W_t)dt + f'(W_t)dW_t + \frac{1}{2}f''(W_t)dt.$$

It is convenient to write this in integrated form,

$$S_t = S_0 + \int_0^t \dot{f}(s, W_s)ds + \int_0^t f'(W_s)dW_s + \int_0^t \frac{1}{2}f''(W_s)ds. \qquad (4.2)$$

Quadratic variation

In order to make sense of a calculus based on Brownian motion, we must find a rigorous mathematical interpretation of the *stochastic integral* (that is, the first integral) on the right hand side of equation (4.2). The key is to study the *quadratic variation* of Brownian motion.

For a typical Brownian path, the 2-variation will be infinite. However, a slightly weaker analogue of 2-variation *does* exist.

Theorem 4.2.1   *Let W_t denote Brownian motion under $\mathbb{P}$ and for a partition π of $[0, T]$ define*

$$S(\pi) = \sum_{j=1}^{N(\pi)} \left|W_{t_j} - W_{t_{j-1}}\right|^2.$$

Let π_n be a sequence of partitions with $\delta(\pi_n) \to 0$. Then

$$\mathbb{E}\left[|S(\pi_n) - T|^2\right] \to 0 \qquad \text{as } n \to \infty. \qquad (4.3)$$

We say that the quadratic variation process of Brownian motion, denoted by $\{[W]_t\}_{t \geq 0}$, is $[W]_t = t$. More generally, we can define the quadratic variation process associated with any bounded continuous martingale.

Definition 4.2.2   *Suppose that $\{M_t\}_{t \geq 0}$ is a bounded continuous $\mathbb{P}$-martingale. The quadratic variation process associated with $\{M\}_{t \geq 0}$ is the process $\{[M]_t\}_{t \geq 0}$ such that for any sequence of partitions π_n of $[0, T]$ with $\delta(\pi_n) \to 0$,*

$$\mathbb{E}\left[\left|\sum_{j=1}^{N(\pi)} \left|M_{t_j} - M_{t_{j-1}}\right|^2 - [M]_T\right|^2\right] \to 0 \qquad \text{as } n \to \infty. \qquad (4.4)$$

Remark: We don't prove it here, but the limit in (4.4) will be independent of the sequence of partitions.   □

Proof of Theorem 4.2.1: We expand the expression inside the expectation in (4.3) and make use of our knowledge of the normal distribution. Let $\{t_{n,j}\}_{j=0}^{N(\pi_n)}$ denote the endpoints of the intervals that make up the partition π_n. First observe that

$$|S(\pi_n) - T|^2 = \left| \sum_{j=1}^{N(\pi_n)} \left\{ |W_{t_{n,j}} - W_{t_{n,j-1}}|^2 - (t_{n,j} - t_{n,j-1}) \right\} \right|^2 .$$

It is convenient to write $\delta_{n,j}$ for $|W_{t_{n,j}} - W_{t_{n,j-1}}|^2 - (t_{n,j} - t_{n,j-1})$. Then

$$|S(\pi_n) - T|^2 = \sum_{j=1}^{N(\pi_n)} \delta_{n,j}^2 + 2 \sum_{j<k} \delta_{n,j} \delta_{n,k}.$$

Note that since Brownian motion has independent increments,

$$\mathbb{E}\left[\delta_{n,j} \delta_{n,k}\right] = \mathbb{E}\left[\delta_{n,j}\right] \mathbb{E}\left[\delta_{n,k}\right] = 0 \qquad \text{if } j \neq k.$$

Also

$$\mathbb{E}[\delta_{n,j}^2] = \mathbb{E}\big[|W_{t_{n,j}} - W_{t_{n,j-1}}|^4 \\ - 2|W_{t_{n,j}} - W_{t_{n,j-1}}|^2 (t_{n,j} - t_{n,j-1}) + (t_{n,j} - t_{n,j-1})^2\big].$$

For a normally distributed random variable, X, with mean zero and variance λ, from Exercise 5 of Chapter 3, $\mathbb{E}[|X|^4] = 3\lambda^2$, so we have

$$\begin{aligned} \mathbb{E}[\delta_{n,j}^2] &= 3\left(t_{n,j} - t_{n,j-1}\right)^2 - 2\left(t_{n,j} - t_{n,j-1}\right)^2 + \left(t_{n,j} - t_{n,j-1}\right)^2 \\ &= 2\left(t_{n,j} - t_{n,j-1}\right)^2 \\ &\leq 2\delta(\pi_n)\left(t_{n,j} - t_{n,j-1}\right). \end{aligned}$$

Summing over j

$$\begin{aligned} \mathbb{E}\left[|S(\pi_n) - T|^2\right] &\leq 2 \sum_{j=1}^{N(\pi_n)} \delta(\pi_n)\left(t_{n,j} - t_{n,j-1}\right) \\ &= 2\delta(\pi_n)T \\ &\to 0 \qquad \text{as } n \to \infty. \end{aligned}$$

□

Integrating
Brownian
motion
against itself

This result is not enough to define the integral $\int_0^T f(s, W_s)dW_s$ in the classical way, but it is enough to allow us to essentially mimic the construction of the (Lebesgue) integral, as limits of integrals of simple functions, at least for functions for which $\int_0^T \mathbb{E}[f^2(s, W_s)]ds < \infty$, provided we only require that the limit exist in *an L^2 sense*. That is, if $\{f^{(n)}\}_{n\geq 1}$ is a sequence of step functions converging to f, then $\int_0^t f(s, W_s)dW_s$ will be a random variable for which

$$\mathbb{E}\left[\left|\int_0^t f(s, W_s)dW_s - \int_0^t f^{(n)}(s, W_s)dW_s\right|^2\right] \to 0 \quad \text{as } n \to \infty.$$

This corresponds to replacing the notion of 2-variance by that of quadratic variation in Definition 4.2.2.

Although the construction of the integral may look familiar, its behaviour is far from familiar. We first illustrate this by defining $\int_0^T W_s dW_s$.

From classical integration theory we are used to the idea that

$$\int_0^T f(s, x_s) dx_s = \lim_{\delta(\pi) \to 0} \sum_{j=0}^{N(\pi)-1} f(t_j, x_{t_j}) \left(x_{t_{j+1}} - x_{t_j} \right). \qquad (4.5)$$

Let us define the stochastic integral in the same way, that is

$$\int_0^T W_s dW_s \triangleq \lim_{\delta(\pi) \to 0} \sum_{j=0}^{N(\pi)-1} W_{t_j} \left(W_{t_{j+1}} - W_{t_j} \right),$$

but now with the caveat that the limit may only exist in the L^2 sense.

Consider again the quantity $S(\pi)$ of Theorem 4.2.1.

$$
\begin{aligned}
S(\pi) &= \sum_{j=1}^{N(\pi)} \left(W_{t_j} - W_{t_{j-1}} \right)^2 \\
&= \sum_{j=1}^{N(\pi)} \left\{ \left(W_{t_j}^2 - W_{t_{j-1}}^2 \right) - 2 W_{t_{j-1}} \left(W_{t_j} - W_{t_{j-1}} \right) \right\} \\
&= W_T^2 - W_0^2 - 2 \sum_{j=0}^{N(\pi)-1} W_{t_j} \left(W_{t_{j+1}} - W_{t_j} \right).
\end{aligned}
$$

The left hand side converges to T as $\delta(\pi) \to 0$ (by Theorem 4.2.1) and so letting $\delta(\pi) \to 0$ and rearranging we obtain

$$\int_0^T W_s dW_s = \frac{1}{2} \left(W_T^2 - W_0^2 - T \right).$$

Remark: Notice that this is *not* what one would have predicted from classical integration theory. The extra term in the stochastic integral arises from $\lim_{\delta(\pi) \to 0} S(\pi)$. □

Defining the integral

In equation (4.5), we use $f(t_j, x_{t_j})$ to approximate the value of f on the interval $(t_j, t_{j+1}]$, but in the classical theory we could equally have taken any point inside the interval in place of t_j and, in the limit, the result would have been the same. In the stochastic theory this is no longer the case. In Exercise 3 you are asked to calculate two further limits:

(a)

$$\lim_{\delta(\pi) \to 0} \sum_{j=0}^{N(\pi)-1} W_{t_{j+1}} \left(W_{t_{j+1}} - W_{t_j} \right),$$

(b)

$$\lim_{\delta(\pi)\to 0} \sum_{j=0}^{N(\pi)-1} \left(\frac{W_{t_j} + W_{t_{j+1}}}{2}\right) \left(W_{t_{j+1}} - W_{t_j}\right).$$

By choosing different points within each subinterval of the partition with which to approximate f over the subinterval we obtain *different* integrals. The *Itô integral* of a function $f(s, W_s)$ with respect to W_s is defined (up to a set of $\mathbb{P}$-probability zero) as

$$\int_0^T f(s, W_s) dW_s = \lim_{\delta(\pi)\to 0} \sum_{j=0}^{N(\pi)-1} f(t_j, W_{t_j}) \left(W_{t_{j+1}} - W_{t_j}\right). \qquad (4.6)$$

The *Stratonovich integral* is defined as

$$\int_0^T f(s, W_s) \circ dW_s = \lim_{\delta(\pi)\to 0} \sum_{j=1}^{N(\pi)} \left(\frac{f(t_j, W_{t_j}) + f(t_{j+1}, W_{t_{j+1}})}{2}\right) \left(W_{t_{j+1}} - W_{t_j}\right).$$

Both limits are to be understood in the L^2 sense. The Stratonovich integral has the advantage from the calculational point of view that the rules of Newtonian calculus hold good; cf. Exercise 8. From a modelling point of view, at least for our purposes, it is the *wrong* choice. To see why, think of what is happening over an infinitesimal time interval. We might be modelling, for example, the value of a portfolio. We readjust our portfolio at the *beginning* of the time interval and its change in value over the infinitesimal tick of the clock is beyond our control. A Stratonovich model would allow us to change our portfolio *now* on the basis of the average of two values depending respectively on the current stock price and prices after the next tick of the clock. We don't have that information when we make our investment decisions.

We are simply reiterating what was said in the discrete world. The composition of our portfolio was *previsible*. We make an analogous definition in continuous time.

Definition 4.2.3   *Given a filtration $\{\mathcal{F}_t\}_{t\geq 0}$, the stochastic process $\{X_t\}_{t\geq 0}$ is $\{\mathcal{F}_t\}_{t\geq 0}$-previsible or $\{\mathcal{F}_t\}_{t\geq 0}$-predictable if X_t is $\mathcal{F}_{t-}$-measurable for all t where*

$$\mathcal{F}_{t-} = \bigcup_{s<t} \mathcal{F}_s.$$

Remark: If $\{X_t\}_{t\geq 0}$ is $\{\mathcal{F}_t\}_{t\geq 0}$-adapted and left continuous (so, in particular, if it is continuous) then it is automatically predictable. $\square$

In our Itô stochastic integrals the integrand will always be predictable.

Integrating   We have evaluated the Itô integral in just one special case, when the integrand is itself
simple   Brownian motion. We now extend our repertoire in the same way as in the classical
functions   setting by first considering the integral of simple functions. Throughout we assume
               that $\{W_t\}_{t\geq 0}$ is a $\mathbb{P}$-Brownian motion generating the filtration $\{\mathcal{F}_t\}_{t\geq 0}$.

Definition 4.2.4   *A simple function is one of the form*

$$f(s, \omega) = \sum_{i=1}^{n} a_i(\omega)\mathbf{1}_{I_i}(s),$$

where

$$I_i = (s_i, s_{i+1}], \quad \bigcup_{i=1}^{n} I_i = (0, T], \quad I_i \cap I_j = \{\emptyset\} \text{ if } i \neq j$$

and, for each $i = 1, \ldots, n$, $a_i : \Omega \to \mathbb{R}$ is an $\mathcal{F}_{s_i}$-measurable random variable with $\mathbb{E}[a_i(\omega)^2] < \infty$.

Remark: We have temporarily abandoned our convention of not mentioning Ω. However, this notation has the advantage of capturing both of the key examples that we have in mind, namely a_i a function of W_{s_i} and a_i a function of $\{W_r\}_{0 \leq r \leq s_i}$. We continue to suppress dependence of $\{W_s\}_{0 \leq s \leq T}$ on ω in our notation.   □

Warning: We have defined simple functions to be $\{\mathcal{F}_t\}_{t \geq 0}$-predictable. Some texts would call such functions *simple predictable functions*.

If f is a simple function, then so is $f(s, \omega)\mathbf{1}_{[0,t]}(s)$. We define

$$\int_0^t f(s, \omega)dW_s = \int f(s, \omega)\mathbf{1}_{[0,t]}(s)dW_s.$$

Following (4.6),

$$\int_0^t f(s, \omega)dW_s \triangleq \sum_{i=1}^{n} a_i(\omega)\mathbf{1}_{[0,t]}(s_i)\left(W_{s_{i+1} \wedge t} - W_{s_i}\right).$$

Now, just as for classical integration theory, for a more general (predictable) function, f, we find a sequence of simple functions $\{f^{(n)}\}_{n \geq 1}$ such that $f^{(n)} \to f$ as $n \to \infty$ and define the integral of f with respect to $\{W_s\}_{0 \leq s \leq t}$ to be $\lim_{n \to \infty} \int f^{(n)}(s, \omega)dW_s$ if this limit exists. This won't work for arbitrary f. The next lemma helps identify a space of functions for which we can reasonably expect a nice limit.

Lemma 4.2.5   *Suppose that f is a simple function; then*

1 *the process*

$$\int_0^t f(s, \omega)dW_s$$

is a continuous $\left(\mathbb{P}, \{\mathcal{F}_t\}_{t \geq 0}\right)$-martingale,

2

$$\mathbb{E}\left[\left(\int_0^t f(s, \omega)dW_s\right)^2\right] = \int_0^t \mathbb{E}[f(s, \omega)^2]ds,$$

3

$$\mathbb{E}\left[\sup_{t\leq T}\left(\int_0^t f(s,\omega)dW_s\right)^2\right] \leq 4\int_0^T \mathbb{E}[f(s,\omega)^2]ds.$$

Remark: The second assertion is the famous *Itô isometry*. It suggests that we should be able to extend our definition of the integral over $[0,t]$ to predictable functions such that $\int_0^t \mathbb{E}[f(s,\omega)^2]ds < \infty$. Moreover, for such functions, all three assertions should remain true. In fact, one can extend the definition a little further, but the integral may then fail to be a martingale and this property will be important to us. □

Before proving Lemma 4.2.5 we quote a famous result of Doob.

Theorem 4.2.6 (Doob's inequality)   *If $\{M_t\}_{0\leq t\leq T}$ is a continuous martingale, then*

$$\mathbb{E}\left[\sup_{0\leq t\leq T} M_t^2\right] \leq 4\mathbb{E}[M_T^2].$$

The proof of this remarkable theorem can be found, for example, in Chung & Williams (1990).

Proof of Proof of Lemma 4.2.5.: The proof of the first assertion is Exercise 5 and the third follows from the second by an application of Doob's inequality, so we confine ourselves to proving the second statement.

We simplify notation by supposing that, in the notation of Definition 4.2.4,

$$f(s,\omega)\mathbf{1}_{[0,t]}(s) = \sum_{i=1}^n a_i(\omega)\mathbf{1}_{I_i}(s)$$

where the intervals I_i are disjoint and $\bigcup_{i=1}^n I_i = (0,t]$. By our definition we have

$$\int_0^t f(s,\omega)dW_s = \sum_{i=1}^n a_i(\omega)\left(W_{s_{i+1}} - W_s\right)$$

and so

$$\mathbb{E}\left[\left(\int_0^t f(s,\omega)dW_s\right)^2\right] = \mathbb{E}\left[\left(\sum_{i=1}^n a_i(\omega)\left(W_{s_{i+1}} - W_{s_i}\right)\right)^2\right]$$

$$= \mathbb{E}\left[\sum_{i=1}^n a_i^2(\omega)\left(W_{s_{i+1}} - W_{s_i}\right)^2\right]$$

$$+ 2\mathbb{E}\left[\sum_{i<j} a_i(\omega)a_j(\omega)\left(W_{s_{i+1}} - W_{s_i}\right)\left(W_{s_{j+1}} - W_{s_j}\right)\right].$$

Suppose that $j > i$; then by the tower property of conditional expectations

$$\mathbb{E}\left[a_i(\omega)a_j(\omega)\left(W_{s_{i+1}} - W_{s_i}\right)\left(W_{s_{j+1}} - W_{s_j}\right)\right]$$
$$= \mathbb{E}\left[a_i(\omega)a_j(\omega)\left(W_{s_{i+1}} - W_{s_i}\right)\mathbb{E}\left[\left(W_{s_{j+1}} - W_{s_j}\right)\middle|\mathcal{F}_{s_j}\right]\right] = 0.$$

Moreover,

$$\mathbb{E}\left[a_i^2(\omega)\left(W_{s_{i+1}} - W_{s_i}\right)^2\right] = \mathbb{E}\left[a_i^2(\omega)\mathbb{E}\left[\left(W_{s_{i+1}} - W_{s_i}\right)^2\middle|\mathcal{F}_{s_i}\right]\right]$$
$$= \mathbb{E}\left[a_i^2(\omega)\right](s_{i+1} - s_i).$$

Substituting we obtain

$$\mathbb{E}\left[\left(\int_0^t f(s,\omega)dW_s\right)^2\right] = \sum_{i=1}^n \mathbb{E}\left[a_i^2(\omega)\right](s_{i+1} - s_i)$$
$$= \int_0^t \mathbb{E}\left[f(s,\omega)^2\right]ds$$

as required.   □

Notation: We write $\mathcal{H}_T$ for the set of functions $f : \mathbb{R}_+ \times \Omega \to \mathbb{R}$ for which $f(t,\omega)$ is $\{\mathcal{F}_t\}_{t\geq 0}$-predictable for $0 \leq t \leq T$ and

$$\int_0^T \mathbb{E}\left[f(s,\omega)^2\right]ds < \infty.$$

Construction of the Itô integral

This will be our class of integrable functions. We proceed as advertised: approximate a general $f \in \mathcal{H}_T$ by a sequence of simple functions, $\{f^{(n)}\}_{n\geq 1}$, and define

$$\int_0^t f(s,\omega)ds \triangleq \lim_{n\to\infty}\int_0^t f^{(n)}(s,\omega)dW_s.$$

The following theorem confirms that this really works.

Theorem 4.2.7   *Suppose that $\{W_t\}_{t\geq 0}$ is a $\mathbb{P}$-Brownian motion and let $\{\mathcal{F}_t\}_{t\geq 0}$ denote its natural filtration. There exists a linear mapping, J, from $\mathcal{H}_T$ to the space of continuous $\left(\mathbb{P}, \{\mathcal{F}_t\}_{t\geq 0}\right)$-martingales defined on $[0, T]$ such that*

1   *if f is simple and $t \leq T$,*

$$J(f)_t = \int_0^t f(s,\omega)dW_s,$$

2   *if $t \leq T$,*

$$\mathbb{E}[J(f)_t^2] = \int_0^t \mathbb{E}\left[f(s,\omega)^2\right]ds,$$

3

$$\mathbb{E}\left[\sup_{0\le t\le T} J(f)_t^2\right] \le 4\int_0^T \mathbb{E}\left[f(s,\omega)^2\right]ds.$$

Proof: By Doob's inequality, the last part follows from the second once we know that $\{J(f)_t\}_{0\le t\le T}$ is a $\mathbb{P}$-martingale. To define J and prove the first two assertions, we follow the approximation procedure outlined above.

Let $\{f^{(n)}\}_{n\ge 1}$ be a sequence of simple functions such that

$$\mathbb{E}\left[\int_0^t \left|f(s,\omega) - f^{(n)}(s,\omega)\right|^2 ds\right] \to 0 \quad \text{as } n\to\infty.$$

Since the difference of two simple functions is simple, by Lemma 4.2.5

$$\mathbb{E}\left[\sup_{0\le t\le T}\left(J(f^{(n)})_t - J(f^{(m)})_t\right)^2\right] \le 4\int_0^T \mathbb{E}\left[\left|f^{(n)}(s,\omega) - f^{(m)}(s,\omega)\right|^2\right]ds$$
$$\to \quad 0 \quad \text{as } n,m\to\infty. \tag{4.7}$$

We now define $J(f)_t$ to be $\lim_{n\to\infty} J(f^{(n)})_t$. From (4.7) the limit exists *uniformly* for $0\le t\le T$, except possibly on a set of $\mathbb{P}$-probability zero, where we set $J(f)_t$ to be identically equal to zero. Moreover,

$$\mathbb{E}\left[J(f)_t^2\right] = \lim_{n\to\infty}\mathbb{E}\left[J(f^{(n)})_t^2\right] = \lim_{n\to\infty}\int_0^t \mathbb{E}\left[f^{(n)}(s,\omega)^2\right]ds = \int_0^t \mathbb{E}\left[f(s,\omega)^2\right]ds.$$

It remains to check the martingale property.

Now by Jensen's inequality (stated in Exercise 16 of Chapter 2), which works equally well for conditional expectations,

$$\mathbb{E}\left[\left|\mathbb{E}\left[J(f^{(n)})_t|\mathcal{F}_s\right] - \mathbb{E}\left[J(f)_t|\mathcal{F}_s\right]\right|^2\right] = \mathbb{E}\left[\mathbb{E}\left[J(f^{(n)})_t - J(f)_t\,\Big|\,\mathcal{F}_s\right]^2\right]$$
$$\le \mathbb{E}\left[\mathbb{E}\left[\left|J(f^{(n)})_t - J(f)_t\right|^2\,\Big|\,\mathcal{F}_s\right]\right]$$
$$= \mathbb{E}\left[\left|J(f^{(n)})_t - J(f)_t\right|^2\right]$$
$$\to \quad 0 \quad \text{as } n\to\infty.$$

So using

$$J(f^{(n)})_s = \mathbb{E}\left[J(f^{(n)})_t\,\Big|\,\mathcal{F}_s\right]$$

and taking limits

$$\mathbb{E}\left[\left(J(f)_s - \mathbb{E}[J(f)_t|\mathcal{F}_s]\right)^2\right] = 0.$$

This implies

$$J(f)_s = \mathbb{E}[J(f)_t|\mathcal{F}_s] \quad \text{with } \mathbb{P}\text{-probability one.}$$

This is almost the martingale property but we want to remove the 'almost surely' qualification. To do this choose a version of $J(f)$ such that the martingale property holds for all $s, t \in \mathbb{Q}$ (we can do this by redefining $J(f)$ on a set of $\mathbb{P}$-measure zero). Since $J(f)$ is a uniform limit of continuous functions (or identically zero) it is continuous and so with this definition the martingale property holds for all s, t as required.   □

Definition 4.2.8   *For $f \in \mathcal{H}_T$, we write*

$$J(f)_t = \int_0^t f(s, \omega)dW_s$$

and call this quantity the Itô stochastic integral *of f with respect to $\{W_t\}_{t \geq 0}$.*

Notice that $J(f)$ really does agree with the prescription (4.6) except possibly on a set of $\mathbb{P}$-probability zero.

Other integrators   We have defined the stochastic integral with respect to Brownian motion. An easy extension is to any process $\{X_t\}_{t \geq 0}$ that can be written as $X_t = W_t + A_t$ where $\{W_t\}_{t \geq 0}$ is Brownian motion and $\{A_t\}_{t \geq 0}$ is a continuous process of bounded variation. In that case we can define the integral with respect to $\{X_t\}_{t \geq 0}$ as the sum of two parts: the integral with respect to the Brownian motion plus that with respect to $\{A_t\}_{t \geq 0}$. The latter exists in the classical sense. We can also replace Brownian motion by other martingales and that is our next goal.

Suppose that $\{M_t\}_{t \geq 0}$ is a continuous $(\mathbb{P}, \{\mathcal{F}_t\}_{t \geq 0})$-martingale with $\mathbb{E}\left[M_t^2\right] < \infty$ for each $t > 0$. By analogy with the Itô integral with respect to Brownian motion, for a simple function

$$f(s, \omega) = \sum_{i=1}^{n} a_i(\omega) 1_{I_i}(s),$$

we define

$$\int f(s, \omega)dM_s \triangleq \sum_{i=1}^{n} a_i(\omega)\left(M_{s_{i+1}} - M_{s_i}\right).$$

Passing to limits we can then define

$$J^M(f)_t = \int_0^t f(s, \omega)dM_s$$

for all $f \in \mathcal{H}_T^M$ where $\mathcal{H}_T^M$ is the set of predictable functions $f : \mathbb{R}_+ \times \mathbb{R} \to \mathbb{R}$ such that

$$\int_0^T \mathbb{E}\left[f(s, \omega)^2\right]d[M]_s < \infty.$$

By redefining $J^M(f)$ to be zero on a set of $\mathbb{P}$-measure zero if necessary, we obtain the following analogue of Theorem 4.2.7.

Theorem 4.2.9   *Assume that $\{M_t\}_{t \geq 0}$ is a bounded continuous $(\mathbb{P}, \{\mathcal{F}_t\}_{t \geq 0})$-martingale with $\mathbb{E}\left[M_t^2\right] < \infty$ for each $t > 0$. Then there exists a linear mapping J^M from $\mathcal{H}_T^M$ to the space of continuous $(\mathbb{P}, \{\mathcal{F}_t\}_{t \geq 0})$-martingales defined on $[0, T]$ such that*

1   *if f is simple, then*

$$J^M(f)_t = \int_0^t f(s, \omega) dM_s = \int f(s, \omega) 1_{[0,t]}(s) dM_s,$$

as defined above,

2   *if $t \leq T$*

$$\mathbb{E}\left[J^M(f)_t^2\right] = \mathbb{E}\left[\int_0^t f(s, \omega)^2 d[M]_s\right],$$

where $\{[M]_t\}_{t\geq 0}$ is the quadratic variation process associated with $\{M_t\}_{t\geq 0}$, and

3

$$\mathbb{E}\left[\sup_{0\leq t\leq T} J^M(f)_t^2\right] \leq 4\mathbb{E}\left[\int_0^T f(s, \omega)^2 d[M]_s\right].$$

Except possibly on a set of $\mathbb{P}$-probability zero,

$$\int_0^t f(s, \omega) dM_s = \lim_{\delta(\pi)\to 0} \sum_{i=0}^{N(\pi)-1} f(s_i, \omega)\left(M_{s_{i+1}} - M_{s_i}\right).$$

We can now extend the definition still further to define the integral with respect to any process $\{X_t\}_{t\geq 0}$ that can be written as $X_t = M_t + A_t$ for a continuous martingale $\{M_t\}_{t\geq 0}$ (with $\mathbb{E}\left[M_t^2\right] < \infty$) and a process $\{A_t\}_{t\geq 0}$ of bounded variation.

We'll exploit this greater generality in proving Lemma 4.2.11, a useful result that tells us that martingales of bounded variation are constant.

Definition 4.2.10   *Suppose that $\{M_t\}_{t\geq 0}$ is a continuous martingale and $\{A_t\}_{t\geq 0}$ is a process of bounded variation; then the process $\{X_t\}_{t\geq 0}$ defined by $X_t = M_t + A_t$ is said to be a* semimartingale.

A continuous semimartingale is any process that can be decomposed in this way. If we insist that $A_0 = 0$ then the decomposition is unique.

Warning: Strictly we should replace 'martingale' by 'local martingale' in Definition 4.2.10. See, for example, Ikeda & Watanabe (1989) or Chung & Williams (1990) for a more general treatment.

Lemma 4.2.11   *Let $\{A_t\}_{0\leq t\leq T}$ be a continuous $(\mathbb{P}, \{\mathcal{F}_t\}_{0\leq t\leq T})$-martingale with $\mathbb{E}\left[A_t^2\right] < \infty$ for each $0 \leq t \leq T$. If $\{A_t\}_{0\leq t\leq T}$ has bounded variation on $[0, T]$ then*

$$\mathbb{P}\left[A_t = A_0, \forall t \in [0, T]\right] = 1.$$

Proof: Let $\hat{A}_t = A_t - A_0$. Since $\{\hat{A}_t\}_{0\leq t\leq T}$ is a continuous process of bounded variation we can define the integral $\int_0^t \hat{A}_s d\hat{A}_s$ in the classical way, and by the

Fundamental Theorem of Calculus,

$$\hat{A}_t^2 - \hat{A}_0^2 = \hat{A}_t^2 = 2 \int_0^t \hat{A}_s d\hat{A}_s.$$

The integral will be the same whether viewed as a classical or as a stochastic integral and so Theorem 4.2.9 tells us that it is a martingale and hence

$$\mathbb{E}\left[2 \int_0^t \hat{A}_s d\hat{A}_s \right] = 0.$$

Thus $\mathbb{E}[\hat{A}_t^2] = 0$ for all t and so by continuity of $\{\hat{A}_t\}_{0 \leq t \leq T}$, $\mathbb{P}[\hat{A}_t = 0, \forall t \in [0, T]] = 1$ as required. $\square$

4.3   Itô's formula

Having made some sense of the stochastic integral, we are now in a position to establish some of the rules of Itô stochastic calculus. We begin with the chain rule and some of its ramifications.

The stochastic chain rule

Theorem 4.3.1 (Itô's formula)   *For f such that the partial derivatives $\frac{\partial f}{\partial t}$, $\frac{\partial f}{\partial x}$ and $\frac{\partial^2 f}{\partial x^2}$ exist and are continuous and $\frac{\partial f}{\partial x} \in \mathcal{H}$, almost surely for each t we have*

$$f(t, W_t) - f(0, W_0) = \int_0^t \frac{\partial f}{\partial x}(s, W_s)dW_s + \int_0^t \frac{\partial f}{\partial s}(s, W_s)ds$$
$$+ \frac{1}{2} \int_0^t \frac{\partial^2 f}{\partial x^2}(s, W_s)ds.$$

Notation: Often one writes Itô's formula in differential notation as

$$df(t, W_t) = f'(t, W_t)dW_t + \dot{f}(t, W_t)dt + \frac{1}{2}f''(t, W_t)dt.$$

Outline of proof: To avoid too many cumbersome formulae, suppose that $\dot{f}_t \equiv 0$. (The proof extends without difficulty to the general case.) The formula then becomes

$$f(W_t) - f(W_0) = \int_0^t \frac{\partial f}{\partial x}(W_s)dW_s + \frac{1}{2} \int_0^t \frac{\partial^2 f}{\partial x^2}(W_s)ds. \qquad (4.8)$$

Let π be a partition of $[0, t]$ and as usual write t_i, t_{i+1} for the endpoints of a generic interval. Then

$$f(W_t) - f(W_0) = \sum_{j=0}^{N(\pi)-1} \left(f(W_{t_{j+1}}) - f(W_{t_j}) \right).$$

We apply Taylor's Theorem on each interval of the partition.

$$f(W_t) - f(W_0) = \sum_{j=0}^{N(\pi)-1} f'(W_{t_j})\left(W_{t_{j+1}} - W_{t_j}\right) + \frac{1}{2}\sum_{j=0}^{N(\pi)-1} f''(\xi_j)\left(W_{t_{j+1}} - W_{t_j}\right)^2,$$

for some points $\xi_j \in \left[W_{t_j} \wedge W_{t_{j+1}}, W_{t_j} \vee W_{t_{j+1}}\right]$. By continuity of the Brownian path, we can write this as

$$f(W_t) - f(W_0) = \sum_{j=0}^{N(\pi)-1} f'(W_{t_j})(W_{t_{j+1}} - W_{t_j}) + \frac{1}{2}\sum_{j=0}^{N(\pi)-1} f''(W_{\eta_j})(W_{t_{j+1}} - W_{t_j})^2,$$

where $\eta_j \in [t_j, t_{j+1}]$. We rewrite the second term as

$$\frac{1}{2}\sum_{j=0}^{N(\pi)-1} \left(f''(W_{t_j}) + \epsilon_j\right)\left(W_{t_{j+1}} - W_{t_j}\right)^2,$$

where $\epsilon_j = f''(W_{\eta_j}) - f''(W_{t_j})$.

A special case: Suppose that $\frac{\partial^2 f}{\partial x^2}$ is bounded. Then, for each fixed T, since $r \mapsto \frac{\partial^2 f}{\partial x^2}(W_r)$ is uniformly continuous on $[0, T]$, $\sup_j \epsilon_j \to 0$ as the mesh of the partition tends to zero. Now we mimic our proof of Theorem 4.2.1.

$$\mathbb{E}\left[\left|\sum_{j=0}^{N(\pi)-1} f''(W_{t_j})\left(\left(W_{t_{j+1}} - W_{t_j}\right)^2 - (t_{j+1} - t_j)\right)\right|^2\right]$$

$$= \mathbb{E}\left[\sum_{j=0}^{N(\pi)-1} (f''(W_{t_j}))^2 \left(\left(W_{t_{j+1}} - W_{t_j}\right)^2 - (t_{j+1} - t_j)\right)^2\right]$$

$$+ 2\mathbb{E}\left[\sum_{0 \le i < j \le N(\pi)-1} f''(W_{t_i})f''(W_{t_j})\left(\left(W_{t_{i+1}} - W_{t_i}\right)^2 - (t_{i+1} - t_i)\right)\right.$$

$$\left. \times \left(\left(W_{t_{j+1}} - W_{t_j}\right)^2 - (t_{j+1} - t_j)\right)\right]. \tag{4.9}$$

Exactly as before, conditioning on $\mathcal{F}_{t_j}$ and using the tower property of conditional expectations, coupled now with boundedness of $\frac{\partial^2 f}{\partial x^2}$, shows that the right hand side of (4.9) tends to zero as the mesh of the partition tends to zero.

Finally, using that

$$\sum_{j=0}^{N(\pi)-1} \frac{\partial f}{\partial x}(W_{t_j})\left(W_{t_{j+1}} - W_{t_j}\right) \to \int_0^t \frac{\partial f}{\partial x}(W_s)dW_s,$$

and exploiting continuity, we see that if $\frac{\partial^2 f}{\partial x^2}$ is bounded, the formula (4.8) holds except possibly on a set of $\mathbb{P}$-measure zero.

The general case: In order to drop the assumption that $\frac{\partial^2 f}{\partial x^2}$ is bounded, we can use what is called a 'localising sequence'. Let

$$\tau_n = \inf \{t \geq 0 : |W_t| > n\}.$$

Then replacing $\{W_s\}_{s \geq 0}$ by $\{W_{s \wedge \tau_n}\}_{s \geq 0}$, since $\frac{\partial^2 f}{\partial x^2}$ is continuous, $\left\{\frac{\partial^2 f}{\partial x^2}(W_{s \wedge \tau_n})\right\}_{s \geq 0}$ is uniformly bounded. Our proof goes through with $\{W_s\}_{s \geq 0}$ replaced by $\{W_{s \wedge \tau_n}\}_{s \geq 0}$ throughout. (Note that we need the fact that τ_n is a stopping time to make this work.) The full result then follows by letting $n \to \infty$. $\square$

For full details of the proof of Itô's formula, see, for example, Ikeda & Watanabe (1989) or Chung & Williams (1990).

Example 4.3.2   *Use Itô's formula to compute* $\mathbb{E}[W_t^6]$.

Solution: Let us define $\{Z_t\}_{t \geq 0}$ by $Z_t = W_t^6$. Then by Itô's formula

$$dZ_t = 6W_t^5 dW_t + 15W_t^4 dt,$$

and, of course, $Z_0 = 0$. In integrated form,

$$Z_t - Z_0 = \int_0^t 6W_s^5 dW_s + \int_0^t 15W_s^4 ds.$$

The expectation of the stochastic integral vanishes (by the martingale property) and so

$$\mathbb{E}[Z_t] = \int_0^t 15\mathbb{E}[W_s^4] ds.$$

Now from Exercise 5 of Chapter 3 (or Exercise 9 of this chapter), $\mathbb{E}[W_s^4] = 3s^2$, and so substituting

$$\mathbb{E}[W_t^6] = \mathbb{E}[Z_t] = 15\int_0^t 3s^2 ds = 15t^3.$$

$\square$

Geometric Brownian motion

The basic reference model for stock prices in continuous time is *geometric Brownian motion*, defined by

$$S_t = S_0 \exp(\nu t + \sigma W_t),$$ (4.10)

where, as usual, $\{W_t\}_{t \geq 0}$ is a standard $\mathbb{P}$-Brownian motion. Applying Itô's formula,

$$dS_t = \sigma S_t dW_t + \left(\nu + \frac{1}{2}\sigma^2\right) S_t dt.$$

This expression is called the *stochastic differential equation* for S_t. It is common to write such symbolic equations even though it is the *integral* equation that makes sense.

Lemma 4.3.3   *Writing $\mu = \nu + \sigma^2/2$, the geometric Brownian motion process defined above is a $\left(\mathbb{P}, \{\mathcal{F}_t\}_{t\geq 0}\right)$-martingale if and only if $\mu = 0$. Moreover*

$$\mathbb{E}[S_t] = S_0 \exp(\mu t).$$

Proof: Writing the stochastic differential equation for geometric Brownian motion in integrated form,

$$S_t = S_0 + \int_0^t \left(\nu + \frac{1}{2}\sigma^2\right) S_s ds + \int_0^t \sigma S_s dW_s. \tag{4.11}$$

Notice that $\{S_t\}_{t\geq 0}$ is a $\left(\mathbb{P}, \{\mathcal{F}_t\}_{t\geq 0}\right)$-semimartingale. Proving that it is a martingale if and only if $\mu = 0$ amounts to proving uniqueness of the decomposition of a semimartingale into a martingale and a bounded variation process in this special case.

Suppose $\mu = 0$: The classical integral in (4.11) then vanishes. Since $\{S_t\}_{t\geq 0}$ inherits continuity from the Brownian motion and by (4.10) it is adapted, by the remark after Definition 4.2.3 it is predictable and so by Theorem 4.2.7 the stochastic integral in (4.11) is a martingale.

Suppose that $\{S_t\}_{t\geq 0}$ is a $(\mathbb{P}, \{\mathcal{F}_t\}_{t\geq 0})$-martingale: Since the difference of two martingales (with respect to the same filtration) is again a martingale, we see that $\{A_t\}_{t\geq 0}$ defined by

$$A_t = S_t - S_0 - \int_0^t \sigma S_s dW_s = \int_0^t \mu S_s ds$$

is a martingale. This classical integral has bounded variation and so by Lemma 4.2.11 with probability one it is equal to $A_0 = 0$. Since $S_s > 0$ for all s, it follows that $\mu = 0$.

The expectation: To verify the second claim, we take expectations in (4.11) and use once again that the stochastic integral term is a mean zero martingale to obtain

$$\mathbb{E}[S_t] - S_0 = \mathbb{E}\left[\int_0^t \mu S_s ds\right] = \int_0^t \mu \mathbb{E}[S_s] ds.$$

(The interchange of time integral and expectation is justified by classical integration theory.) Solving this integral equation gives

$$\mathbb{E}[S_t] = S_0 \exp(\mu t)$$

Itô's formula for geometric Brownian motion

as required.   □

It is convenient to have a version of Itô's formula that allows us to work directly with $\{S_t\}_{t\geq 0}$ (so that we can write down a stochastic differential equation for $f(t, S_t)$). We now know how to make our original heuristic calculations rigorous, so with a clear conscience we proceed as follows:

$$f(t + \delta t, S_{t+\delta t}) - f(t, S_t) \approx \dot{f}(t, S_t)\delta t + f'(t, S_t)(S_{t+\delta t} - S_t)$$
$$+ \frac{1}{2}f''(t, S_t)(S_{t+\delta t} - S_t)^2$$
$$\approx \dot{f}(t, S_t)dt + f'(t, S_t)dS_t$$
$$+ \frac{1}{2}f''(t, S_t)\{\sigma^2 S_t^2 dW_t^2 + \mu^2 S_t^2 dt^2 + 2\sigma \mu S_t^2 dW_t dt\}.$$

That is

$$df(t, S_t) = \dot{f}(t, S_t)dt + f'(t, S_t)dS_t + \frac{1}{2}f''(t, S_t)\sigma^2 S_t^2 dt,$$

where we have used the multiplication table

$\times$	dW_t	dt
dW_t	dt	0
dt	0	0

Writing this version of Itô's formula in integrated form gives then

$$f(t, S_t) - f(0, S_0) = \int_0^t \frac{\partial f}{\partial u}(u, S_u)du + \int_0^t \frac{\partial f}{\partial x}(u, S_u)dS_u$$
$$+ \frac{1}{2}\int_0^t \frac{\partial^2 f}{\partial x^2}(u, S_u)\sigma^2 S_u^2 du$$
$$= \int_0^t \frac{\partial f}{\partial u}(u, S_u)du + \int_0^t \frac{\partial f}{\partial x}(u, S_u)\sigma S_u dW_u$$
$$+ \int_0^t \frac{\partial f}{\partial x}(u, S_u)\mu S_u du + \frac{1}{2}\int_0^t \frac{\partial^2 f}{\partial x^2}(u, S_u)\sigma^2 S_u^2 du.$$

Warning: Be aware that the stochastic integral with respect to $\{S_t\}_{t\geq 0}$ will not be a martingale with respect to the probability $\mathbb{P}$ under which $\{W_t\}_{t\geq 0}$ is a martingale except in the special case when $\{S_t\}_{t\geq 0}$ is a $\mathbb{P}$-martingale, that is when $\mu = 0$. To actually *calculate* it is often wise to separate the martingale part by expanding the 'stochastic' integral as in the last line.

Example 4.3.4   *Suppose that $\{S_t\}_{t\geq 0}$ is a geometric Brownian motion satisfying*

$$dS_t = \mu S_t dt + \sigma S_t dW_t, \qquad (4.12)$$

where $\{W_t\}_{t\geq 0}$ is standard Brownian motion under $\mathbb{P}$. Calculate $\mathbb{E}\left[S_t^n\right]$ for $n \in \mathbb{N}$.

Solution: From our calculations above,

$$S_t = \exp(\nu t + \sigma W_t)$$

where $\nu = \mu - \sigma^2/2$. Thus $\{S_t^n\}_{t\geq 0}$ is also a geometric Brownian motion with parameters $\nu^{(n)} = n\nu$ and $\sigma^{(n)} = n\sigma$. By Lemma 4.3.3 we then have

$$\mathbb{E}\left[S_t^n\right] = S_0^n \exp\left(\left(n\nu + \frac{1}{2}n^2\sigma^2\right)t\right) = S_0^n \exp\left(\left(n\mu + \frac{1}{2}n(n-1)\sigma^2\right)t\right).$$

(4.13)

To gain some more practice with stochastic calculus, suppose that we did not know how to express the solution to the stochastic differential equation (4.12) explicitly as a function of $\{W_t\}_{t\geq 0}$. An alternative approach to calculating $\mathbb{E}\left[S_t^n\right]$, which we now sketch, is to apply Itô's lemma.

$$\begin{aligned} d\left(S_t^n\right) &= nS_t^{n-1}dS_t + \frac{1}{2}n(n-1)S_t^{n-2}\sigma^2 S_t^2 dt \\ &= S_t^n\left(n\mu + \frac{1}{2}n(n-1)\sigma^2\right)dt + n\sigma S_t^n dW_t. \end{aligned}$$

Writing this equation in integrated form and taking expectations yields

$$\mathbb{E}\left[S_t^n\right] - \mathbb{E}\left[S_0^n\right] = \int_0^t \left(n\mu + \frac{1}{2}n(n-1)\sigma^2\right)\mathbb{E}\left[S_s^n\right]ds.$$

This leads us once again to the expression (4.13).   □

Lévy's characterisation of Brownian motion

The Itô formula provides a quick route to a useful characterisation of Brownian motion due to Lévy. We have seen that Brownian motion is a martingale. It is useful to be able to identify when a martingale is in fact Brownian motion.

Theorem 4.3.5   *Let $\{W_t\}_{0\leq t\leq T}$ be a continuous $(\mathbb{P}, \{\mathcal{F}_t\}_{0\leq t\leq T})$-martingale such that $W_0 = 0$ and $[W]_t = t$ for $0 \leq t \leq T$. Then $\{W_t\}_{0\leq t\leq T}$ is a $(\mathbb{P}, \{\mathcal{F}_t\}_{0\leq t\leq T})$-Brownian motion.*

Proof: We must check that for any $0 \leq s < t \leq T$, $W_t - W_s$ is normally distributed with mean zero and variance $t - s$ and is independent of $\mathcal{F}_s$.

Let

$$M_t^\theta \triangleq \exp\left(\theta W_t - \frac{\theta^2}{2}t\right).$$

Applying Itô's formula we see that $\{M_t\}_{0\leq t\leq T}$ is a $(\mathbb{P}, \{\mathcal{F}_t\}_{0\leq t\leq T})$-martingale and so for $0 \leq s \leq t \leq T$,

$$\mathbb{E}\left[\left.\frac{M_t}{M_s}\right| \mathcal{F}_s\right] = 1.$$

Substituting and rearranging gives

$$\mathbb{E}\left[\exp\left(\theta\left(W_t - W_s\right)\right)|\mathcal{F}_s\right] = \exp\left(\frac{1}{2}(t-s)\theta^2\right).$$

Since the normal distribution is characterised by its moment generating function, the result follows.   □

Stochastic
differential
equations

Let us return to processes of the form $Z_t = f(t, W_t)$. Using Itô's formula

$$dZ_t = \left(\dot{f}(t, W_t) + \frac{1}{2} f''(t, W_t) \right) dt + f'(t, W_t) dW_t.$$

Suppose that f is invertible; then we may write this as

$$dZ_t = \mu(t, Z_t) dt + \sigma(t, Z_t) dW_t. \tag{4.14}$$

Definition 4.3.6   *An equation of the form (4.14) for some deterministic functions* $\mu(t, x)$ *and* $\sigma(t, x)$ *on* $\mathbb{R}_+ \times \mathbb{R}$ *is called a* stochastic differential equation *for* $\{Z_t\}_{t \geq 0}$.

It is often easier to write down a stochastic differential equation for $\{Z_t\}_{t \geq 0}$ than to produce the function $f(t, x)$ explicitly.

> **Warning:**   Just as for (Newtonian) ordinary differential equations, in general a stochastic differential equation may not have a solution and, even if it does, the solution may not be unique.

If the functions $\mu(t, x)$ and $\sigma(t, x)$ are, for example, bounded and uniformly Lipschitz-continuous in x then a unique solution does exist, but these conditions are certainly not necessary (see, for example, Chung & Williams (1990) and Ikeda & Watanabe (1989) for more details).

Of course, we should really understand equation (4.14) in integrated form:

$$Z_t - Z_0 = \int_0^t \mu(s, Z_s) ds + \int_0^t \sigma(s, Z_s) dW_s.$$

It is left to the reader to justify the following version of the Itô formula.

Theorem 4.3.7   *If* $Z_t = f(t, W_t)$ *satisfies*

$$dZ_t = \sigma(t, Z_t) dW_t + \mu(t, Z_t) dt,$$

and

$$Y_t = g(t, Z_t),$$

for some twice differentiable functions f *and* g, *then*

$$dY_t = \dot{g}(t, Z_t) dt + g'(t, Z_t) dZ_t + \frac{1}{2} g''(t, Z_t) \sigma^2(t, Z_t) dt. \tag{4.15}$$

Remark: Notice that

$$M_t = Z_t - Z_0 - \int_0^t \mu(s, Z_s) ds$$

is a martingale with mean zero. From the Itô isometry, we know that its variance is

$$\mathbb{E}[M_t^2] = \mathbb{E}\left[\int_0^t \sigma(s, Z_s)^2 ds\right].$$

The expression $\sigma^2(t, Z_t)dt$ appearing in (4.15) is just $d[M]_t$ where $\{[M]_t\}_{t \geq 0}$ is the quadratic variation associated with $\{M_t\}_{t \geq 0}$.

If $\{Z_t\}_{t \geq 0}$ is defined by $Z_t = M_t + A_t$ where $\{M_t\}_{t \geq 0}$ is a continuous martingale with $\mathbb{E}\left[M_t^2\right] < \infty$ and $\{A_t\}_{t \geq 0}$ has bounded variation, then setting $Y_t = g(t, Z_t)$ we have

$$dY_t = \dot{g}(t, Z_t)dt + g'(t, Z_t)dZ_t + \frac{1}{2}g''(t, Z_t)d[M]_t.$$

In particular, by applying this to $Y_t = M_t^2$, one shows (Exercise 15) that $M_t^2 - [M]_t$ is a martingale.   □

Solving stochastic differential equations

Even when a stochastic differential equation has a unique solution it is rare to be able to express it in closed form as a function of Brownian motion. However, if this can be done, then Itô's formula provides a route to finding the solution.

Example 4.3.8   *Solve*

$$dX_t = X_t^3 dt - X_t^2 dW_t, \qquad X_0 = 1. \tag{4.16}$$

Solution: If $X_t = f(t, W_t)$, then

$$dX_t = \dot{f}(t, W_t)dt + f'(t, W_t)dW_t + \frac{1}{2}f''(t, W_t)dt,$$

and, substituting in (4.16),

$$dX_t = f(t, W_t)^3 dt - f(t, W_t)^2 dW_t.$$

Equating coefficients we obtain

$$-f(t, W_t)^2 = f'(t, W_t)$$

and

$$\dot{f}(t, W_t) + \frac{1}{2}f''(t, W_t) = f(t, W_t)^3,$$

which yields

$$f(t, x) = \frac{1}{x + c}$$

where c is a constant. Using the initial condition, we find

$$X_t = \frac{1}{W_t + 1}.$$

Notice that this solution blows up in finite time.   □

4.4   Integration by parts and a stochastic Fubini Theorem

We shall need two more rules for manipulating stochastic integrals: the integration by parts formula and a 'stochastic Fubini Theorem'. The first is the product rule of stochastic differentiation, the second is used to justify interchange of order of stochastic and classical integrals.

Suppose that we have *two* stochastic differential equations:

$$dY_t = \mu(t, Y_t)dt + \sigma(t, Y_t)dW_t,$$

$$dZ_t = \tilde{\mu}(t, Z_t)dt + \tilde{\sigma}(t, Z_t)dW_t.$$

We assume for now that these equations are driven by the *same* Brownian motion $\{W_t\}_{t \geq 0}$.

Consider the $(\mathbb{P}, \{\mathcal{F}_t\}_{t \geq 0})$-martingales defined by

$$M_t^Y = \int_0^t \sigma(s, Y_s)dW_s \quad \text{and} \quad M_t^Z = \int_0^t \tilde{\sigma}(s, Z_s)dW_s,$$

with associated quadratic variation processes

$$[M^Y]_t = \int_0^t \sigma^2(s, Y_s)ds \quad \text{and} \quad [M^Z]_t = \int_0^t \tilde{\sigma}^2(s, Z_s)ds.$$

Covariation Clearly $\{M_t^Y\}_{t \geq 0}$ and $\{M_t^Z\}_{t \geq 0}$ are not independent. One way of quantifying the dependence between them is through their covariance. Evidently $\{(M_t^Y + M_t^Z)\}_{t \geq 0}$ and $\{(M_t^Y - M_t^Z)\}_{t \geq 0}$ are also $(\mathbb{P}, \{\mathcal{F}_t\}_{t \geq 0})$-martingales with quadratic variation

$$[M^Y \pm M^Z]_t = \int_0^t (\sigma(s, Y_s) \pm \tilde{\sigma}(s, Z_s))^2 \, ds.$$

We're interested in the process $\{M_t^Y M_t^Z\}_{t \geq 0}$. We attack it via polarisation:

$$M_t^Y M_t^Z = \frac{1}{4}\left(\left(M_t^Y + M_t^Z\right)^2 - \left(M_t^Y - M_t^Z\right)^2 \right).$$

Taking expectations,

$$\mathbb{E}\big[M_t^Y M_t^Z\big] = \frac{1}{4}\mathbb{E}\left[\int_0^t (\sigma(s, Y_s) + \tilde{\sigma}(s, Z_s))^2 \, ds - \int_0^t (\sigma(s, Y_s) - \tilde{\sigma}(s, Z_s))^2 \, ds \right]$$

$$= \mathbb{E}\left[\int_0^t \sigma(s, Y_s)\tilde{\sigma}(s, Z_s)ds \right].$$

We write

$$\big[M^Y, M^Z\big]_t = \int_0^t \sigma(s, Y_s)\tilde{\sigma}(s, Z_s)ds.$$

Since for a $(\mathbb{P}, \{\mathcal{F}_t\}_{t \geq 0})$-martingale $\{M_t^2 - [M]_t\}_{t \geq 0}$ is a $(\mathbb{P}, \{\mathcal{F}_t\}_{t \geq 0})$-martingale (Exercise 15), again exploiting polarisation we see that $\{M_t^Y M_t^Z - [M^Y, M^Z]_t\}_{t \geq 0}$ is a $(\mathbb{P}, \{\mathcal{F}_t\}_{t \geq 0})$-martingale.

More generally, we could consider stochastic differential equations driven by *different* (but not necessarily independent) Brownian motions. For example, suppose that

$$dY_t = \mu(t, Y_t)dt + \sigma(t, Y_t)dW_t,$$

$$dZ_t = \tilde{\mu}(t, Z_t)dt + \tilde{\sigma}(t, Z_t)d\tilde{W}_t,$$

where $\mathbb{E}\big[(W_t - W_s)(\tilde{W}_t - \tilde{W}_s)\big] = \rho(t - s)$ for some $0 < \rho < 1$. This means that the Brownian motions driving the two equations are positively correlated, an increase in one tends to be associated with an increase in the other, but they are not identical. We define M_t^Y and M_t^Z exactly as before and again study $M_t^Y M_t^Z$ via polarisation. Writing

$$\big[M_t^Y M_t^Z\big] = \frac{1}{4}\left(\big[M^Y + M^Z\big]_t - \big[M^Y - M^Z\big]_t\right),$$

and using the definition of quadratic variation,

$$\big[M^Y, M^Z\big]_t = \lim_{\delta(\pi)\to 0} \sum_{j=0}^{N(\pi)-1} \left(M_{t_{j+1}}^Y - M_{t_j}^Y\right)\left(M_{t_{j+1}}^Z - M_{t_j}^Z\right).$$

In our example, provided σ and $\tilde{\sigma}$ are continuous say, we have

$$\big[M^Y, M^Z\big]_t = \lim_{\delta(\pi)\to 0} \sum_{j=0}^{N(\pi)-1} \sigma(t_j, Y_{t_j})\tilde{\sigma}(t_j, Z_{t_j})\left(W_{t_{j+1}} - W_{t_j}\right)\left(\tilde{W}_{t_{j+1}} - \tilde{W}_{t_j}\right),$$

and by mimicking the argument of the proof of Theorem 4.2.1 we obtain

$$\big[M^Y, M^Z\big]_t = \int_0^t \sigma(s, Y_s)\tilde{\sigma}(s, Z_s)\rho ds.$$

Definition 4.4.1   *For continuous* $(\mathbb{P}, \{\mathcal{F}_t\}_{t\geq 0})$-*martingales* $\{M_t\}_{t\geq 0}$ *and* $\{N_t\}_{t\geq 0}$

$$[M, N]_t \triangleq \frac{1}{4}\big([M + N]_t - [M - N]_t\big)$$

is called the mutual variation *or* covariation *process of M and N.*

Of course $\{[M, M]_t\}_{t\geq 0}$ is just the quadratic variation process associated with $\{M_t\}_{t\geq 0}$. In the notation of Definition 4.4.1, if we write $\delta(\pi)$ for a generic partition of $[0, t]$ then

$$[M, N]_t = \lim_{\delta(\pi)\to 0} \sum_{j=0}^{N(\pi)-1} \left(M_{t_{j+1}} - M_{t_j}\right)\left(N_{t_{j+1}} - N_{t_j}\right).$$

We now have the technology required for manipulating products of semimartingales.

Theorem 4.4.2 (Integration by parts)   *If* $Y_t = M_t^Y + A_t^Y$ *and* $Z_t = M_t^Z + A_t^Z$, *where* $\{M_t^Y\}_{t\geq 0}$ *and* $\{M_t^Z\}_{t\geq 0}$ *are continuous* $(\mathbb{P}, \{\mathcal{F}_t\}_{t\geq 0})$-*martingales and* $\{A_t^Y\}_{t\geq 0}$ *and* $\{A_t^Z\}_{t\geq 0}$ *are continuous processes of bounded variation, then*

$$d(Y_t Z_t) = Y_t dZ_t + Z_t dY_t + d\big[M^Y, M^Z\big]_t.$$

Proof: We apply the Itô formula to $(Y_t + Z_t)^2$, Y_t^2 and Z_t^2, and subtract the second two from the first to obtain the result.   □

Example 4.4.3   *Suppose that the Sterling price of an asset follows the stochastic differential equation*

$$dS_t = \mu_1 S_t dt + \sigma_1 S_t dW_t$$

and the dollar cost of £1 at time t is E_t where

$$dE_t = \mu_2 E_t dt + \sigma_2 E_t d\tilde{W}_t.$$

Here $\{W_t\}_{t\geq 0}$ and $\{\tilde{W}_t\}_{t\geq 0}$ are $\mathbb{P}$-Brownian motions with

$$\mathbb{E}\left[(W_t - W_s)(\tilde{W}_t - \tilde{W}_s)\right] = \rho(t - s)$$

for some constant $\rho > 0$.

If the riskless interest rate is r in the UK and s in the USA, find the stochastic differential equation for the discounted asset price in the Sterling and dollar markets respectively. For what values of the parameters are the discounted asset prices martingales in each market?

Solution: In the Sterling market, write $\{\tilde{S}\}_{t\geq 0}$ for the discounted stock price. That is $\tilde{S}_t = e^{-rt} S_t$. Since the function e^{-rt} has bounded variation, our integration by parts formula gives

$$
\begin{aligned}
d\tilde{S}_t &= -re^{-rt} S_t dt + e^{-rt} dS_t \\
&= (\mu_1 - r)\, \tilde{S}_t dt + \sigma_1 \tilde{S}_t dW_t.
\end{aligned}
$$

In Sterling markets, this is the stochastic differential equation governing the discounted asset price. The solution is a martingale if and only if $\mu_1 = r$.

Let us write $\{X_t\}_{t\geq 0}$ for the dollar price of the asset. Then $X_t = E_t S_t$ and, again by integration by parts,

$$
\begin{aligned}
dX_t &= E_t dS_t + S_t dE_t + \sigma_1 \sigma_2 E_t S_t \rho dt \\
&= (\mu_1 + \mu_2 + \rho\sigma_1\sigma_2)\, X_t dt + \sigma_1 X_t dW_t + \sigma_2 X_t d\tilde{W}_t.
\end{aligned}
$$

The discounted asset price in the dollar market, denoted by $\{\tilde{X}_t\}_{t\geq 0}$, then follows

$$d\tilde{X}_t = (\mu_1 + \mu_2 + \rho\sigma_1\sigma_2 - s)\, \tilde{X}_t dt + \sigma_1 \tilde{X}_t dW_t + \sigma_2 \tilde{X}_t d\tilde{W}_t.$$

The discounted price in the dollar market is a martingale if and only if

$$\mu_1 + \mu_2 + \rho\sigma_1\sigma_2 - s = 0.$$

Notice that it is perfectly possible for the discounted asset price to be a martingale in one market but not the other. It is important to keep this in mind when valuing options in the foreign exchange market (see §5.3) or when valuing quantos (see §7.2).   □

A stochastic   We finish this section with one more useful result. This is a 'stochastic Fubini
Fubini   Theorem' that will allow us to interchange the order of a stochastic and a classical
Theorem   integral. We shall need this result in valuation of certain path-dependent exotic
                  options in Exercise 23 of Chapter 6.

We state the theorem in a very special form that will be sufficient for our needs. For a more general version see, for example, Ikeda & Watanabe (1989).

Theorem 4.4.4   Let $(\Omega, \mathcal{F}, \{\mathcal{F}_t\}_{t\geq 0}, \mathbb{P})$ be a filtered probability space and let $\{M_t\}_{t\geq 0}$ be a continuous $(\mathbb{P}, \{\mathcal{F}_t\}_{t\geq 0})$-martingale with $M_0 = 0$. Suppose that $\Phi(t, r, \omega) : \mathbb{R}_+ \times \mathbb{R}_+ \times \Omega \to \mathbb{R}$ is a bounded $\{\mathcal{F}_t\}_{t\geq 0}$-predictable random variable. Then for each fixed $T > 0$,

$$\int_0^t \int_{\mathbb{R}} \Phi(s, r, \omega) \, 1_{[0,T]}(r) dr dM_s = \int_{\mathbb{R}} \int_0^t \Phi(s, r, \omega) \, 1_{[0,T]}(r) dM_s dr.$$

Example 4.4.5   Suppose that $\{W_t\}_{t\geq 0}$ is a $(\mathbb{P}, \{\mathcal{F}_t\}_{t\geq 0})$-Brownian motion. Evaluate the mean and variance of

$$Y_t \triangleq \int_0^t W_r dr.$$

Solution: The classical Fubini Theorem tells us that

$$\mathbb{E}[Y_t] = \mathbb{E}\left[\int_0^t W_r dr\right] = \int_0^t \mathbb{E}[W_r] dr = 0.$$

The difficulty is to calculate $\mathbb{E}[Y_t^2]$ and this is where we exploit our stochastic Fubini Theorem with $\Phi(s, r, \omega) = 1_{[0,r]}(s)$.

$$\begin{aligned}
Y_t = \int_0^t W_r dr &= \int_0^t \int_0^r 1 dW_s dr \\
&= \int_0^t \int_s^t 1 dr dW_s \quad \text{(Fubini's Theorem)} \\
&= \int_0^t (t - s) dW_s.
\end{aligned}$$

Now using the Itô isometry we can calculate $\mathbb{E}[Y_t^2]$ to be

$$\mathbb{E}[Y_t^2] = \mathbb{E}\left[\left(\int_0^t (t - s) dW_s\right)^2\right] = \int_0^t (t - s)^2 ds = \frac{1}{3}t^3.$$

□

4.5   The Girsanov Theorem

In order to price and hedge in the Black–Scholes framework we shall need two fundamental results. The first will allow us to change probability measure so that the discounted asset prices are martingales. Recall that in our discrete time world, once we had such a *martingale measure*, the pricing of options was reduced to calculating expectations under that measure. In the continuous world it will no longer be possible to find the martingale measure by linear algebra. Nonetheless, before stating the continuous time result, we revert to our binomial trees for guidance.

Changing
probability
on a
binomial
tree

We use the notation of Chapter 2. Suppose that, under the probability measure $\mathbb{P}$, if the value of an asset at time $i\delta t$ is known to be S_i then its value at time $(i+1)\delta t$ is $S_i u$ with probability p and it is $S_i d$ with probability $1 - p$.

As we saw in Chapter 2, if we let $\mathbb{Q}$ be the probability measure under which the probability of an up jump is $q = (1-d)/(u-d)$ and of a down jump is $(u-1)/(u-d)$, then the process $\{S_i\}_{0 \leq i \leq N}$ is a $\mathbb{Q}$-martingale.

We can regard the measure $\mathbb{Q}$ as a *reweighting* of the measure $\mathbb{P}$. For example, consider a path $S_0, S_1, \ldots, S_i$ through the tree. Its probability under $\mathbb{P}$ is $p^{N(i)}(1-p)^{i-N(i)}$, where $N(i)$ is the number of up jumps that the path makes. Under $\mathbb{Q}$ its probability is $L_i p^{N(i)}(1-p)^{i-N(i)}$ where

$$L_i = \left(\frac{q}{p}\right)^{N(i)}\left(\frac{1-q}{1-p}\right)^{i-N(i)}.$$

Evidently L_i depends on the path that the stochastic process takes through the tree and can itself be thought of as a stochastic process adapted to the filtration $\{\mathcal{F}_i\}_{1 \leq i \leq N}$. Moreover, L_i/L_{i-1} is q/p if $S_i/S_{i-1} = u$ and is $(1-q)/(1-p)$ if $S_i/S_{i-1} = d$, so that

$$\mathbb{E}^{\mathbb{P}}\left[L_i | \mathcal{F}_{i-1}\right] = L_{i-1}\left(p\frac{q}{p} + (1-p)\frac{1-q}{1-p}\right) = L_{i-1}.$$

In other words, $\{L_i\}_{0 \leq i \leq N}$ is a $\left(\mathbb{P}, \{\mathcal{F}_i\}_{1 \leq i \leq N}\right)$-martingale with $\mathbb{E}[L_i] = L_0 = 1$.

If we wish to calculate the expected value of a claim in the $\mathbb{Q}$-measure, it is given by

$$\mathbb{E}^{\mathbb{Q}}[C] = \mathbb{E}^{\mathbb{P}}[L_N C].$$

Notation: We have obtained the *Radon–Nikodym derivative* of $\mathbb{Q}$ with respect to $\mathbb{P}$. It is customary to write

$$L_i = \frac{d\mathbb{Q}}{d\mathbb{P}}\Big|_{\mathcal{F}_i}.$$

Change of
measure in
the
continuous
world

We have shown that the process of changing to the martingale measure can be viewed as a reweighting of the probabilities of paths under our original measure $\mathbb{P}$ according to a positive, mean one, $\mathbb{P}$-martingale. This procedure of reweighting according to a positive martingale can be extended to the continuous setting. Our aim now is to investigate the effect of such a reweighting on the distribution of the $\mathbb{P}$-Brownian motion. Later this will enable us to choose the right reweighting so that under the new measure obtained in this way the discounted stock price is a martingale.

Theorem 4.5.1 (Girsanov's Theorem)   *Suppose that $\{W_t\}_{t \geq 0}$ is a $\mathbb{P}$-Brownian motion with the natural filtration $\{\mathcal{F}_t\}_{t \geq 0}$ and that $\{\theta_t\}_{t \geq 0}$ is an $\{\mathcal{F}_t\}_{t \geq 0}$-adapted process such that*

$$\mathbb{E}\left[\exp\left(\frac{1}{2}\int_0^T \theta_t^2 dt\right)\right] < \infty.$$

Define

$$L_t = \exp\left(-\int_0^t \theta_s dW_s - \frac{1}{2}\int_0^t \theta_s^2 ds\right)$$

and let $\mathbb{P}^{(L)}$ be the probability measure defined by

$$\mathbb{P}^{(L)}[A] = \int_A L_t(\omega)\mathbb{P}(d\omega).$$

Then under the probability measure $\mathbb{P}^{(L)}$, the process $\{W_t^{(L)}\}_{0 \leq t \leq T}$, defined by

$$W_t^{(L)} = W_t + \int_0^t \theta_s ds,$$

is a standard Brownian motion.

Notation: We write

$$\left.\frac{d\mathbb{P}^{(L)}}{d\mathbb{P}}\right|_{\mathcal{F}_t} = L_t.$$

(L_t is the *Radon–Nikodym derivative* of $\mathbb{P}^{(L)}$ with respect to $\mathbb{P}$.)

Remarks:

1   The condition

$$\mathbb{E}\left[\exp\left(\frac{1}{2}\int_0^T \theta_t^2 dt\right)\right] < \infty,$$

known as Novikov's condition, is enough to guarantee that $\{L_t\}_{t \geq 0}$ is a $(\mathbb{P}, \{\mathcal{F}_t\}_{t \geq 0})$-martingale. Since L_t is clearly positive and has expectation one, $\mathbb{P}^{(L)}$ really does define a probability measure.

2   Just as in the discrete world, two probability measures are equivalent if they have the same sets of probability zero. Evidently $\mathbb{P}$ and $\mathbb{P}^{(L)}$ are equivalent.

3   If we wish to calculate an expectation with respect to $\mathbb{P}^{(L)}$ we have

$$\mathbb{E}^{\mathbb{P}^{(L)}}[\phi_t] = \mathbb{E}[\phi_t L_t].$$

More generally,

$$\mathbb{E}^{\mathbb{P}^{(L)}}[\phi_t | \mathcal{F}_s] = \mathbb{E}^{\mathbb{P}}\left[\phi_t \frac{L_t}{L_s}\bigg| \mathcal{F}_s\right].$$

This will be fundamental in option pricing.   □

Outline of proof of theorem: We have already said that $\{L_t\}_{t\geq 0}$ is a $(\mathbb{P}, \{\mathcal{F}_t\}_{t\geq 0})$-martingale. We don't prove this in full, but we find supporting evidence by finding the stochastic differential equation satisfied by L_t. We do this in two stages. First, define

$$Z_t = -\int_0^t \theta_s dW_s - \frac{1}{2}\int_0^t \theta_s^2 ds.$$

Then

$$dZ_t = -\theta_t dW_t - \frac{1}{2}\theta_t^2 dt.$$

Now we apply Itô's formula to $L_t = \exp(Z_t)$.

$$
\begin{aligned}
dL_t &= \exp(Z_t)dZ_t + \frac{1}{2}\exp(Z_t)\theta_t^2 dt \\
&= -\theta_t \exp(Z_t)dW_t = -\theta_t L_t dW_t.
\end{aligned}
$$

Now we use our integration by parts formula of Theorem 4.4.2 to find the stochastic differential equation satisfied by $W_t^{(L)}L_t$. Since

$$
\begin{aligned}
dW_t^{(L)} &= dW_t + \theta_t dt, \\
d(W_t^{(L)}L_t) &= W_t^{(L)}dL_t + L_t dW_t^{(L)} + d[M^{W^{(L)}}, M^L]_t \\
&= W_t^{(L)}dL_t + L_t dW_t + L_t \theta_t dt - \theta_t L_t dt \\
&= \left(L_t - \theta_t L_t W_t^{(L)}\right)dW_t.
\end{aligned}
$$

Granted enough boundedness (which is guaranteed by our assumptions), $\{W_t^{(L)}L_t\}_{t\geq 0}$ is then a $\mathbb{P}$-*martingale* and has expectation zero. Thus, under the measure $\mathbb{P}^{(L)}$, $\{W_t^{(L)}\}_{t\geq 0}$ is a martingale.

We proved in Theorem 4.2.1 that with $\mathbb{P}$-probability one, the quadratic variation of $\{W_t\}_{t\geq 0}$ is given by $[W]_t = t$. The probability measures $\mathbb{P}$ and $\mathbb{P}^{(L)}$ are equivalent and so have the same sets of probability one. Therefore $\{W_t^{(L)}\}_{t\geq 0}$ also has quadratic variation given by $[W^{(L)}]_t = t$ with $\mathbb{P}^{(L)}$-probability one. Finally, by Lévy's characterisation of Brownian motion (Theorem 4.3.5) we have that $\{W_t^{(L)}\}_{t\geq 0}$ is a $\mathbb{P}^{(L)}$-Brownian motion as required.   □

We now try this in practice.

Example 4.5.2   *Let $\{X_t\}_{t\geq 0}$ be the drifting Brownian motion process*

$$X_t = \sigma W_t + \mu t,$$

where $\{W_t\}_{t\geq 0}$ is a $\mathbb{P}$-Brownian motion and σ and μ are constants. Find a measure under which $\{X_t\}_{t\geq 0}$ is a martingale.

Solution: Taking $\theta = \mu/\sigma$, under $\mathbb{P}^{(L)}$ of Theorem 4.5.1 we have that $W_t^{(L)} = W_t + \mu t/\sigma$ is a Brownian motion, and $X_t = \sigma W_t^{(L)}$ is then a scaled Brownian motion.

Notice that, for example,

$$\mathbb{E}^{\mathbb{P}}\left[X_t^2\right] = \mathbb{E}^{\mathbb{P}}\left[\sigma^2 W_t^2 + 2\sigma\mu t W_t + \mu^2 t^2\right] = \sigma^2 t + \mu^2 t^2,$$

whereas

$$\mathbb{E}^{\mathbb{P}^{(L)}}\left[X_t^2\right] = \mathbb{E}^{\mathbb{P}^{(L)}}\left[\sigma^2 (W_t^{(L)})^2\right] = \sigma^2 t.$$

□

4.6   The Brownian Martingale Representation Theorem

Let $\left(\Omega, \mathcal{F}, \{\mathcal{F}_t\}_{t\geq 0}, \mathbb{P}\right)$ be a filtered probability space and let $\{W_t\}_{t\geq 0}$ be a $(\mathbb{P}, \{\mathcal{F}_t\}_{t\geq 0})$-Brownian motion. We have seen that if $f(t, \omega)$ is an $\{\mathcal{F}_t\}_{t\geq 0}$-predictable random variable and $\mathbb{E}\left[f^2(t, \omega)\right] < \infty$ for each $t \geq 0$, then

$$M_t \triangleq \int_0^t f(s, \omega) dW_s$$

is a $\left(\mathbb{P}, \{\mathcal{F}_t\}_{t\geq 0}\right)$-martingale. It is natural to ask if there are any others.

Just as in the discrete world the binomial representation theorem allowed us to represent martingales as 'discrete stochastic integrals' so here the Brownian martingale representation theorem tells us that all (nice) $\left(\mathbb{P}, \{\mathcal{F}_t\}_{t\geq 0}\right)$-martingales can be represented as Itô integrals. This result is also sometimes called the *predictable representation property*. It will be the key to *hedging* in our continuous world.

Definition 4.6.1   A $\left(\mathbb{P}, \{\mathcal{F}_t\}_{t\geq 0}\right)$-*martingale* $\{M_t\}_{t\geq 0}$ *is said to be square-integrable if*

$$\mathbb{E}\left[|M_t|^2\right] < \infty \text{ for each } t > 0.$$

Theorem 4.6.2 (Brownian Martingale Representation Theorem)   *Let $\{\mathcal{F}_t\}_{t\geq 0}$ denote the natural filtration of the $\mathbb{P}$-Brownian motion $\{W_t\}_{t\geq 0}$. Let $\{M_t\}_{t\geq 0}$ be a square-integrable $\left(\mathbb{P}, \{\mathcal{F}_t\}_{t\geq 0}\right)$-martingale. Then there exists an $\{\mathcal{F}_t\}_{t\geq 0}$-predictable process $\{\theta_t\}_{t\geq 0}$ such that with $\mathbb{P}$-probability one,*

$$M_t = M_0 + \int_0^t \theta_s dW_s.$$

Outline of proof: We restrict our attention to $t \in [0, T]$ for some fixed T. The first step is to show that any $F \in \mathcal{F}_T$ for which $\mathbb{E}\left[F^2\right] < \infty$ can be represented as

$$F = \mathbb{E}[F] + \int_0^T \theta_s dW_s \qquad\qquad (4.17)$$

for some predictable process $\{\theta_s\}_{0\leq s \leq T}$. Write $\mathcal{G}$ for the linear space of such F that *can* be represented in this way. For any $F \in \mathcal{G}$,

$$\mathbb{E}\left[F^2\right] = \mathbb{E}[F]^2 + \mathbb{E}\left[\int_0^T \theta_s^2 ds\right]. \qquad\qquad (4.18)$$

This guarantees that if we take a sequence of random variables $\{F_n\}_{n\geq 1}$ in $\mathcal{G}$ for which

$$\mathbb{E}\left[(F_n - F_m)^2\right] \to 0 \quad \text{as } n, m \to \infty,$$

then they converge to a limit that also lies in $\mathcal{G}$. Now by Itô's formula, for any simple function

$$f(s) = \sum_{i=1}^{n} a_i(\omega) 1_{(t_{i-1}, t_i]}(s),$$

if we define

$$\mathcal{E}_t^f \triangleq \exp\left(\int_0^t f(s) dW_s - \frac{1}{2} \int_0^t f(s)^2 ds\right),$$

then

$$\mathcal{E}_t^f = 1 + \int_0^t f(s) \mathcal{E}_s^f dW_s.$$

In other words $\mathcal{E}_T^f \in \mathcal{G}$. We now approximate any $F \in \mathcal{F}_T$ for which $\mathbb{E}\left[F^2\right] < \infty$ by linear combinations of the $\mathcal{E}_T^f$ to see that all such F are in $\mathcal{G}$ and so can be represented as in (4.17). The identity (4.18) guarantees that if the representation holds with two different predictable processes, $\{\theta_s\}_{0\leq s\leq T}$ and $\{\psi_s\}_{0\leq s\leq T}$ say, then

$$\mathbb{E}\left[\int_0^T (\theta_s - \psi_s)^2 ds\right] = 0.$$

Now we replace $F \in \mathcal{F}_T$ by the martingale $\{M_t\}_{0\leq t\leq T}$ to complete the proof. This step is elementary. Since $M_t = \mathbb{E}[M_T | \mathcal{F}_t]$, applying the representation to M_T and then taking (conditional) expectations of both sides we obtain

$$\begin{aligned}
M_t = \mathbb{E}[M_T | \mathcal{F}_t] &= \mathbb{E}[M_T] + \mathbb{E}\left[\int_0^T \theta_s dW_s \,\middle|\, \mathcal{F}_t\right] \\
&= \mathbb{E}[M_T] + \int_0^t \theta_s dW_s.
\end{aligned}$$

$\square$

For full details of this proof see Revuz & Yor (1998).

Remarks:

1 The Martingale Representation Theorem tells us that such an $\{\mathcal{F}_t\}_{t\geq 0}$-predictable process $\{\theta_t\}_{t\geq 0}$ *exists*. Unfortunately, unlike the Binomial Representation Theorem, the proof is not constructive. When we call upon it in hedging options in Chapter 5, we are going to have to work harder to actually produce an *explicit* expression for the predictable process.

2 Notice that the quadratic variation of the martingale $\{M_t\}_{t\geq 0}$ satisfies $d[M]_t = \theta_t^2 dt$. If we have two Brownian martingales, $\{M_t^{(1)}\}_{t\geq 0}$ and $\{M_t^{(2)}\}_{t\geq 0}$, then provided $d[M^{(i)}]_t/dt$ is non-vanishing for $i = 1, 2$, the Martingale Representation Theorem tells us that each is a locally scaled version of the other. $\square$

4.7 Why geometric Brownian motion?

We now have the main results in place that will allow us to price and hedge in stock market models based on Brownian motion. Other than suggesting that the paths of stock prices in an arbitrage-free world should be rough, we have thus far provided no justification for such models. In this short section we use Lévy's characterisation of Brownian motion to motivate the basic reference model in mathematical finance: geometric Brownian motion.

We begin by sketching Bachelier's argument for the Brownian motion model. Bachelier argued that stock markets cannot have any consistent bias in favour of either buyers or sellers:

'L'espérance mathématique du spéculateur est nulle'.

This is almost the martingale property. Assuming the stock price process to have the Markov property, he introduced the transition density

$$\mathbb{P}\big[S_t \in [y, y + dy] \big| S_s = x\big] \triangleq p\,(s, t; x, y)\,dy.$$

If the dynamics are homogeneous in space and time, then $p(s, t; x, y) = q(t-s, y - x)$ for some function q. Bachelier then 'derived' what is now known as the Chapman–Kolmogorov equation for q and showed that this is solved by the probability density function of Brownian motion.

Bachelier's argument is not rigorous, but from Lévy's characterisation of Brownian motion we know that if the stock price process is a martingale under $\mathbb{P}$ whose increments have stationary conditional variance then the stock price process must be Brownian motion under $\mathbb{P}$. It is remarkable that Bachelier's argument pre-dates Einstein's famous work on Brownian motion and, of course, Wiener's rigorous construction of the process.

Although we would not take issue with the mathematical conclusions of Bachelier's analysis, we have already discarded Brownian motion as a model. A modern approach makes different assumptions, but we need not completely abandon Bachelier's argument. His key assumption was that the increments of the stock price process were stationary. Suppose that instead we assume that the *relative* increments, $(S_t - S_s)/S_s$, measuring the *returns* are stationary. Taking logarithms, the process $\{\log S_t\}_{t\geq 0}$ should have stationary increments. We don't know that $\{\log S_t\}_{t\geq 0}$ is a martingale, so this time we can only deduce that this is Brownian motion plus a constant drift. This gives

$$dS_t = \mu S_t dt + \sigma S_t dW_t,$$

where $\{W_t\}_{t\geq 0}$ is a $\mathbb{P}$-Brownian motion and μ and σ are constants. This is the geometric Brownian motion model, originally championed by Samuelson (1965).

4.8 The Feynman–Kac representation

Our probabilistic approach to pricing options will result in a price expressed as the discounted expected value of a claim with respect to a probability measure under which the discounted stock price is a martingale. In the simple case of European calls and puts we'll be able to find an explicit expression for the price. However, for more complicated claims numerical methods have to be brought to bear. One approach is to revert to our binomial tree model. Another is to express the price as the solution to a partial differential equation and employ, for example, finite difference methods. In fact for European options the binomial method amounts to a finite difference method for solving the Black–Scholes partial differential equation. We refer to Wilmott, Howison & Dewynne (1995) for an account of the numerical methods. Here we simply make the connection between the partial differential equation approach and the probabilistic approach to pricing derivatives.

Solving
pde's proba-
bilistically
    The fact that the price can be expressed as the solution to a partial differential equation is a consequence of the deep connection between stochastic differential equations and certain parabolic partial differential equations.

Theorem 4.8.1 (Feynman–Kac stochastic representation)   *Assume that the function F solves the boundary value problem*

$$\frac{\partial F}{\partial t}(t, x) + \mu(t, x)\frac{\partial F}{\partial x}(t, x) + \frac{1}{2}\sigma^2(t, x)\frac{\partial^2 F}{\partial x^2}(t, x) = 0, \qquad 0 \le t \le T,$$

$$F(T, x) = \Phi(x).$$
(4.19)

Define $\{X_t\}_{0 \le t \le T}$ to be the solution of the stochastic differential equation

$$dX_t = \mu(t, X_t)dt + \sigma(t, X_t)dW_t, \qquad 0 \le t \le T,$$

where $\{W_t\}_{t \ge 0}$ is standard Brownian motion under the measure $\mathbb{P}$. If

$$\int_0^T \mathbb{E}\left[\left(\sigma(t, X_t)\frac{\partial F}{\partial x}(t, X_t)\right)^2\right] ds < \infty,$$
(4.20)

then

$$F(t, x) = \mathbb{E}^{\mathbb{P}}\left[\Phi(X_T)|\, X_t = x\right].$$

Proof: We apply Itô's formula to $\{F(s, X_s)\}_{t \le s \le T}$.

$$F(T, X_T) = F(t, X_t)$$
$$+ \int_t^T \left\{\frac{\partial F}{\partial s}(s, X_s) + \mu(s, X_s)\frac{\partial F}{\partial x}(s, X_s) + \frac{1}{2}\sigma^2(s, X_s)\frac{\partial^2 F}{\partial x^2}(s, X_s)\right\} ds$$
$$+ \int_t^T \sigma(s, X_s)\frac{\partial F}{\partial x}(s, X_s)dW_s.$$
(4.21)

Now using assumption (4.20) and Theorem 4.2.7

$$\mathbb{E}\left[\int_t^T \sigma(s, X_s)\frac{\partial F}{\partial x}(s, X_s)dW_s \,\middle|\, X_t = x\right] = 0.$$

Moreover, since F satisfies (4.19), the deterministic integral on the right hand side of (4.21) vanishes, so, taking expectations,

$$\mathbb{E}[F(T, X_T)| X_t = x] = F(t, x),$$

and substituting $F(T, X_T) = \Phi(X_T)$ gives the required result.   □

Example 4.8.2   *Solve*

$$\frac{\partial F}{\partial t} + \frac{1}{2}\frac{\partial^2 F}{\partial x^2} = 0, \tag{4.22}$$
$$F(T, x) = \Phi(x).$$

Solution: The corresponding stochastic differential equation is

$$dX_t = dW_t$$

so, by the Feynman–Kac representation,

$$F(t, x) = \mathbb{E}[\Phi(W_T)| W_t = x].$$

In fact we knew this already. In §3.1 we wrote down the transition density of Brownian motion as

$$p(t, x, y) = \frac{1}{\sqrt{2\pi t}}\exp\left(-\frac{(x-y)^2}{2t}\right). \tag{4.23}$$

This gives
$$\mathbb{E}[\Phi(W_T)| W_t = x] = \int p(T-t, x, y)\,\Phi(y)dy.$$

To check that this really is the solution, differentiate and use the fact that $p(t, x, y)$ given by (4.23) is the fundamental solution to the equation

$$\frac{\partial u}{\partial t} = \frac{1}{2}\frac{\partial^2 u}{\partial x^2},$$

to obtain (4.22).   □

Kolmogorov equations   We can use the Feynman–Kac representation to find the partial differential equation solved by the transition densities of solutions to other stochastic differential equations.

Suppose that

$$dX_t = \mu(t, X_t)dt + \sigma(t, X_t)dW_t. \tag{4.24}$$

For any set B let

$$p_B(t, x; T, y) \triangleq \mathbb{P}[X_T \in B| X_t = x] = \mathbb{E}[\mathbf{1}_B(X_T)| X_t = x].$$

By the Feynman–Kac representation (subject to the integrability condition (4.20)) this solves

$$\frac{\partial p_B}{\partial t}(t, x; T, y) + A p_B(t, x; T, y) \; = \; 0, \qquad (4.25)$$

$$p_B(T, x) \; = \; 1_B(x),$$

where

$$A f(t, x) = \mu(t, x)\frac{\partial f}{\partial x}(t, x) + \frac{1}{2}\sigma^2(t, x)\frac{\partial^2 f}{\partial x^2}(t, x).$$

Writing $|B|$ for the Lebesgue measure of the set B, the transition density of the process $\{X_s\}_{s \geq 0}$ is given by

$$p(t, x; T, y) \triangleq \lim_{|B| \to 0}\frac{1}{|B|}\mathbb{P}[X_T \in B \,|\, X_t = x].$$

Since the equation (4.25) is linear, we have proved the following lemma.

Lemma 4.8.3   *Subject to satisfying (4.20), the transition density of the solution $\{X_s\}_{s \geq 0}$ to the stochastic differential equation (4.24) solves*

$$\frac{\partial p}{\partial t}(t, x; T, y) + A p(t, x; T, y) \; = \; 0, \qquad (4.26)$$

$$p(t, x; T, y) \; \to \; \delta_y(x) \quad as\ t \to T.$$

Equation (4.26) is known as the *Kolmogorov backward equation* (it operates on the 'backward in time' variables (t, x)). The operator A is called the *infinitesimal generator* of the process $\{X_s\}_{s \geq 0}$.

We can also obtain an equation acting on the *forward* variables (T, y).

Lemma 4.8.4   *In the above notation,*

$$\frac{\partial p}{\partial T}(t, x; T, y) = A^* p(t, x; T, y) \qquad (4.27)$$

where

$$A^* f(T, y) = -\frac{\partial}{\partial y}\left(\mu(T, y) f(T, y)\right) + \frac{1}{2}\frac{\partial^2}{\partial y^2}\left(\sigma^2(t, Y) f(T, y)\right).$$

Heuristic explanation: We don't prove this, but we provide some justification. By the Markov property of the process $\{X_t\}_{t \geq 0}$, for any $T > r > t$

$$p(t, x; T, y) = \int p(t, x; r, z) p(r, z; T, y) dz.$$

Differentiating with respect to r and using (4.26),

$$\int_{-\infty}^{\infty}\left\{\frac{\partial}{\partial r}p(t, x; r, z) p(r, z; T, y) + p(t, x; r, z) A p(r, z; T, y)\right\} dz = 0.$$

Now integrate the second term by parts to obtain

$$\int_{-\infty}^{\infty} \left\{ \frac{\partial}{\partial r} p(t, x; r, z) - A^* p(t, x; r, z) \right\} p(r, z; T, y) dz = 0.$$

This holds for all $T > r$, which, if $p(r, z; T, y)$ provides a sufficiently rich class of functions as we vary T, implies the result. □

Equation (4.27) is the *Kolmogorov forward equation* of the process $\{X_s\}_{s\geq0}$.

Example 4.8.5   *Find the forward and backward Kolmogorov equations for geometric Brownian motion.*

Solution: The stochastic differential equation is

$$dS_t = \mu S_t dt + \sigma S_t dW_t.$$

Substituting in our formula for the forward equation we obtain

$$\frac{\partial p}{\partial T}(t, x; T, y) = \frac{1}{2}\sigma^2 \frac{\partial^2}{\partial y^2}\left(y^2 p(t, x; T, y)\right) - \mu \frac{\partial}{\partial y}(yp(t, x; T, y)),$$

and the backward equation is

$$\frac{\partial p}{\partial t}(t, x; T, y) = -\frac{1}{2}\sigma^2 x^2 \frac{\partial^2 p}{\partial x^2}(t, x; T, y) - \mu x \frac{\partial p}{\partial x}(t, x; T, y).$$

The transition density for the process is the lognormal density given by

$$p(t, x; T, y) = \frac{1}{\sigma y\sqrt{2\pi(T-t)}} \exp\left(-\frac{\left(\log(y/x) - (\mu - \frac{1}{2}\sigma^2)(T-t)\right)^2}{2\sigma^2(T-t)}\right).$$

□

Example 4.8.6   *Suppose that $\{X_t\}_{t\geq0}$ solves*

$$dX_t = \mu(t, X_t)dt + \sigma(t, X_t)dW_t,$$

where $\{W_t\}_{t\geq0}$ is a $\mathbb{P}$-Brownian motion. For $k : \mathbb{R}_+ \times \mathbb{R} \to \mathbb{R}$ and $\Phi : \mathbb{R} \to \mathbb{R}$ given deterministic functions, find the partial differential equation satisfied by the function

$$F(t, x) \triangleq \mathbb{E}\left[\exp\left(-\int_t^T k(s, X_s)ds\right)\Phi(X_T)\middle| X_t = x\right],$$

for $0 \leq t \leq T$.

Solution: Evidently $F(T, x) = \Phi(x)$. By analogy with the proof of the Feynman–Kac representation, it is tempting to examine the dynamics of

$$Z_s = \exp\left(-\int_t^s k(u, X_u)du\right) F(s, X_s).$$

Notice that for this choice of $\{Z_s\}_{t \leq s \leq T}$ we have that if $X_t = x$

$$Z_t = F(t, x) = \mathbb{E}[Z_T | X_t = x].$$

Thus the partial differential equation satisfied by $F(t, x)$ is that for which $\{Z_t\}_{0 \leq t \leq T}$ is a martingale.

Our strategy now is to find the stochastic differential equation satisfied by $\{Z_s\}_{t \leq s \leq T}$. We proceed in two stages. Remember that t is now fixed and we vary s. First notice that

$$d\left(\exp\left(-\int_t^s k(u, X_u)du\right)\right) = -k(s, X_s)\exp\left(-\int_t^s k(u, X_u)du\right)ds$$

and by Itô's formula

$$
\begin{aligned}
dF(s, X_s) &= \frac{\partial F}{\partial s}(s, X_s)ds + \frac{\partial F}{\partial x}(s, X_s)dX_s + \frac{1}{2}\frac{\partial^2 F}{\partial x^2}(s, X_s)\sigma^2(s, X_s)ds \\
&= \left\{\frac{\partial F}{\partial s}(s, X_s) + \mu(s, X_s)\frac{\partial F}{\partial x}(s, X_s) + \frac{1}{2}\sigma^2(s, X_s)\frac{\partial^2 F}{\partial x^2}(s, X_s)\right\}ds \\
&\quad + \sigma(s, X_s)\frac{\partial F}{\partial x}(s, X_s)dW_s.
\end{aligned}
$$

Now using our integration by parts formula we have that

$$
\begin{aligned}
dZ_s &= \exp\left(-\int_t^s k(u, X_u)du\right) \\
&\quad \times \left\{\left\{-k(s, X_s)F(s, X_s) + \frac{\partial F}{\partial s}(s, X_s) + \mu(s, X_s)\frac{\partial F}{\partial x}(s, X_s) + \frac{1}{2}\sigma^2(s, X_s)\frac{\partial^2 F}{\partial x^2}\right\}ds \right. \\
&\quad \left. + \sigma(s, X_s)\frac{\partial F}{\partial x}(s, X_s)dW_s\right\}.
\end{aligned}
$$

We can now read off the solution: in order for $\{Z_s\}_{t \leq s \leq T}$ to be a martingale, F must satisfy

$$\frac{\partial F}{\partial s}(s, x) + \mu(s, x)\frac{\partial F}{\partial x}(s, x) + \frac{1}{2}\sigma^2(s, x)\frac{\partial^2 F}{\partial x^2}(s, x) - k(s, x)F(s, x) = 0.$$

$\square$

Exercises

1   Let $\{\mathcal{F}_t\}_{t \geq 0}$ denote the natural filtration associated to a standard $\mathbb{P}$-Brownian motion $\{W_t\}_{t \geq 0}$. Define the process $\{S_t\}_{t \geq 0}$ by $S_t = f(t, W_t)$. What equation must f satisfy if S_t is to be a $(\mathbb{P}, \{\mathcal{F}_t\}_{t \geq 0})$-martingale? Use your answer to check that

$$S_t = \exp(vt + \sigma W_t)$$

is a martingale if $v + \frac{1}{2}\sigma^2 = 0$ (cf. Lemma 4.3.3).

2 A function, f, is said to be *Lipschitz-continuous* on $[0, T]$ if there exists a constant $C > 0$ such that for any $t, t' \in [0, T]$

$$|f(t) - f(t')| < C|t - t'|.$$

Show that a Lipschitz-continuous function has bounded variation and zero 2-variation over $[0, T]$.

3 Let $\{W_t\}_{t \geq 0}$ denote a standard Brownian motion under $\mathbb{P}$. For a partition π of $[0, T]$, write $\delta(\pi)$ for the mesh of the partition and $0 = t_0 < t_1 < t_2 < \cdots < t_{N(\pi)} = T$ for the endpoints of the intervals of the partition. Calculate

(a)
$$\lim_{\delta(\pi) \to 0} \sum_{0}^{N(\pi)-1} W_{t_{j+1}} \left(W_{t_{j+1}} - W_{t_j} \right),$$

(b)
$$\int_0^T W_s \circ dW_s \triangleq \lim_{\delta(\pi) \to 0} \sum_{0}^{N(\pi)-1} \frac{1}{2} \left(W_{t_{j+1}} + W_{t_j} \right) \left(W_{t_{j+1}} - W_{t_j} \right).$$

This is the *Stratonovich integral* of $\{W_s\}_{s \geq 0}$ with respect to itself over $[0, T]$.

4 Suppose that the martingale $\{M_t\}_{0 \leq t \leq T}$ has bounded quadratic variation and $\{A_t\}_{0 \leq t \leq T}$ is Lipschitz-continuous. Let $S_t = M_t + A_t$. By analogy with Definition 4.2.2, we define the quadratic variation of $\{S_t\}_{0 \leq t \leq T}$ over $[0, T]$ to be the random variable $[S]_T$ such that for any sequence of partitions $\{\pi_n\}_{n \geq 1}$ of $[0, T]$ with $\delta(\pi_n) \to 0$ as $n \to \infty$,

$$\mathbb{E}\left[\left| \sum_{j=1}^{N(\pi)} |S_{t_j} - S_{t_{j-1}}|^2 - [S]_T \right|^2 \right] \to 0 \quad \text{as } n \to \infty.$$

Show that $[S]_T = [M]_T$.

5 If f is a simple function and $\{W_t\}_{t \geq 0}$ is a $\mathbb{P}$-Brownian motion, prove that the process $\{M_t\}_{t \geq 0}$ given by the Itô integral

$$M_t = \int_0^t f(s, W_s) dW_s$$

is a $(\mathbb{P}, \{\mathcal{F}_t^W\}_{t \geq 0})$-martingale.

6 Verify that if $\{W_t\}_{t \geq 0}$ is a $\mathbb{P}$-Brownian motion

$$\mathbb{E}\left[\left(\int_0^t W_s dW_s \right)^2 \right] = \int_0^t \mathbb{E}\left[W_s^2 \right] ds.$$

(If you need the moment-generating function of W_t, you may assume the result of Exercise 10.)

7   As usual, $\{W_t\}_{t\geq 0}$ denotes standard Brownian motion under $\mathbb{P}$. Use Itô's formula to write down stochastic differential equations for the following quantities.

(a) $Y_t = W_t^3$,
(b) $Y_t = \exp\left(\sigma W_t - \frac{1}{2}\sigma^2 t\right)$,
(c) $Y_t = t W_t$.

Which are $(\mathbb{P}, \{\mathcal{F}_t^W\}_{t\geq 0})$-martingales?

8   Use a heuristic argument based on a Taylor expansion to check that for Stratonovich stochastic calculus the chain rule takes the form of the classical (Newtonian) one.

9   Mimic the calculation of Example 4.3.2 to show that if $\{W_t\}_{t\geq 0}$ is standard Brownian motion under the measure $\mathbb{P}$, then $\mathbb{E}\left[W_t^4\right] = 3t^2$.

10   Let $\{W_t\}_{t\geq 0}$ denote Brownian motion under $\mathbb{P}$ and define $Z_t = \exp(\alpha W_t)$. Use Itô's formula to write down a stochastic differential equation for Z_t. Hence find an ordinary (deterministic) differential equation for $m(t) \triangleq \mathbb{E}[Z_t]$, and solve to show that

$$\mathbb{E}\left[\exp(\alpha W_t)\right] = \exp\left(\frac{\alpha^2}{2}t\right).$$

11   *The Ornstein–Uhlenbeck process* Let $\{W_t\}_{t\geq 0}$ denote standard Brownian motion under $\mathbb{P}$. The Ornstein–Uhlenbeck process, $\{X_t\}_{t\geq 0}$, is the unique solution to *Langevin's equation*,

$$dX_t = -\alpha X_t dt + dW_t, \qquad X_0 = x.$$

This equation was originally introduced as a simple idealised model for the velocity of a particle suspended in a liquid. In finance it is a special case of the *Vasicek model* of interest rates (see Exercise 19). Verify that

$$X_t = e^{-\alpha t}x + e^{-\alpha t}\int_0^t e^{\alpha s}dW_s,$$

and use this expression to calculate the mean and variance of X_t.

12   The *Cox–Ingersoll–Ross model* of interest rates assumes that the interest rate, r, is not deterministic, but satisfies the stochastic differential equation

$$dr_t = (\alpha - \beta r_t)dt + \sigma\sqrt{r_t}dW_t,$$

where $\{W_t\}_{t\geq 0}$ is standard $\mathbb{P}$-Brownian motion. This process is known as a *squared Bessel* process. Find the stochastic differential equation followed by $\{\sqrt{r_t}\}_{t\geq 0}$ in the case $\alpha = 0$.

Suppose that $\{u(t)\}_{t\geq 0}$ satisfies the ordinary differential equation

$$\frac{du}{dt}(t) = -\beta u(t) - \frac{\sigma^2}{2}u(t)^2, \qquad u(0) = \theta,$$

for some constant $\theta > 0$. Fix $T > 0$. Assuming still that $\alpha = 0$, for $0 \le t \le T$ find the differential equation satisfied by

$$\mathbb{E}\left[\exp\left(-u(T-t)r_t\right)\right].$$

Hence calculate the mean and variance of r_T and $\mathbb{P}[r_T = 0]$.

13   The *Black–Karasinski model* of interest rates is

$$dr_t = \sigma_t r_t dW_t + \left(\theta_t + \frac{1}{2}\sigma_t^2 - \alpha_t \log r_t\right) r_t dt,$$

where $\{W_t\}_{t\ge 0}$ is a standard $\mathbb{P}$-Brownian motion and σ_t, θ_t and α_t are deterministic functions of time. In the special case where σ, θ and α are constants, find r_t as a function of $\int_0^t e^{\alpha s} dW_s$.

14   Suppose that, under the measure $\mathbb{P}$,

$$dS_t = \sigma S_t dW_t,$$

where $\{W_t\}_{t\ge 0}$ is a $\mathbb{P}$-Brownian motion. Find the mean and variance of

$$Y_t \triangleq \int_0^t S_u du.$$

15   Suppose that $\{M_t\}_{t\ge 0}$ is a continuous $(\mathbb{P}, \{\mathcal{F}_t\}_{t\ge 0})$-martingale with $\mathbb{E}\left[M_t^2\right]$ finite for all $t \ge 0$. Writing $\{[M]_t\}_{t\ge 0}$ for the associated quadratic variation process, show that $M_t^2 - [M]_t$ is a $(\mathbb{P}, \{\mathcal{F}_t\}_{t\ge 0})$-martingale.

16   Suppose that under the probability measure $\mathbb{P}$, $\{X_t\}_{t\ge 0}$ is a Brownian motion with constant drift μ. Find a measure $\mathbb{P}^*$, equivalent to $\mathbb{P}$, under which $\{X_t\}_{t\ge 0}$ is a Brownian motion with drift ν.

17   Let $\{\mathcal{F}_t\}_{t\ge 0}$ be the natural filtration associated with a $\mathbb{P}$-Brownian motion, $\{W_t\}_{t\ge 0}$. Show that if X is an $\mathcal{F}_T$-measurable random variable with $\mathbb{E}[|X|] < \infty$ and $\mathbb{P}^*$ is a probability measure equivalent to $\mathbb{P}$, then the process

$$M_t \triangleq \mathbb{E}^{\mathbb{P}^*}\left[X|\mathcal{F}_t\right]$$

is a $(\mathbb{P}^*, \{\mathcal{F}_t\}_{0\le t\le T})$-martingale.

18   Use the Feynman–Kac stochastic representation formula to solve

$$\frac{\partial F}{\partial t}(t, x) + \frac{1}{2}\sigma^2 \frac{\partial^2 F}{\partial x^2}(t, x) = 0,$$

subject to the terminal value condition

$$F(T, x) = x^4.$$

19 Suppose that the interest rate, r, is not deterministic, but is itself a random process, $\{r_t\}_{t\geq 0}$. In the *Vasicek model*, $\{r_t\}_{t\geq 0}$ is assumed to be a solution of the stochastic differential equation

$$dr_t = (b - ar_t)dt + \sigma dW_t,$$

where, as usual, $\{W_t\}_{t\geq 0}$ is standard $\mathbb{P}$-Brownian motion.

Find the Kolmogorov backward and forward differential equations satisfied by the probability density function of the process. What is the distribution of r_t as $t \to \infty$?

20 Suppose that $v(t, x)$ solves

$$\frac{\partial v}{\partial t}(t, x) + \frac{1}{2}\sigma^2 x^2 \frac{\partial^2 v}{\partial x^2}(t, x) - rv(t, x) = 0, \quad 0 \leq t \leq T.$$

Show that for any constant θ,

$$v_\theta(t, x) \triangleq \frac{x}{\theta} v\left(t, \frac{\theta^2}{x}\right)$$

is another solution. Use the Feynman–Kac stochastic representation to find a probabilistic interpretation of this result.

21 Suppose that for $0 \leq s \leq T$,

$$dX_s = \mu(s, X_s)ds + \sigma(s, X_s)dW_s, \quad X_t = x,$$

where $\{W_s\}_{t \leq s \leq T}$ is a $\mathbb{P}$-Brownian motion, and let $k, \Phi : \mathbb{R} \to \mathbb{R}$ be given deterministic functions. Find the partial differential equation satisfied by

$$F(t, x) = \mathbb{E}[\Phi(X_T)| X_t = x] + \int_t^T \mathbb{E}[k(X_s)| X_t = x]ds.$$

5 The Black–Scholes model

Summary

We now, finally, have all the tools that we need for pricing and hedging in the continuous time world of Black and Scholes. We shall begin with the most basic setting, in which our market has just two securities: a cash bond and a risky asset whose price is modelled by a geometric Brownian motion.

In §5.1 we prove the Fundamental Theorem of Asset Pricing in this framework. In line with our analysis in the discrete world, this provides an explicit formula for the price of a derivative as the discounted expected payoff under the martingale measure. Just as in the discrete setting, we shall see that there are three steps to replication. In §5.2 we put this into action for European options. For simple calls and puts, the expectation that gives the price of the claim can be evaluated. We also obtain an explicit expression for the stock and bond holding in the replicating portfolio, via an application of the Feynman–Kac representation.

The rest of the book consists of increasing the complexity of the derivative contracts and of the market models. Before embarking on this programme, we relax the financial assumptions that we have made within the basic Black–Scholes framework. The risky asset that we have specified has a very simplistic financial side. We have assumed that it can be held without additional cost or benefit and that it can be freely traded at the quoted price. Even leaving aside the issues of transaction costs and illiquidity, not much of the financial market is like that. Foreign exchange involves two assets that pay interest, equities pay dividends and bonds pay coupons. In §5.3–§5.5 we see how to apply the Black–Scholes technology in these more sophisticated financial settings. Finally, in §5.6, we characterise tradable assets within a given market and we define the market price of risk.

5.1 The basic Black–Scholes model

In this section we provide a rigorous derivation of the Black–Scholes pricing formula obtained in §2.6. As in Chapter 2, our market consists of just two securities. The first is our old friend the cash bond, $\{B_t\}_{t\geq0}$. We retain (for now) our assumption that the risk-free interest rate is constant, so that if $B_0 = 1$ then $B_t = e^{rt}$. The second security in our market is a risky asset whose price at time t we denote by S_t. In this, our basic reference model, we suppose that $\{S_t\}_{t\geq 0}$ is a geometric Brownian motion, that is it solves

$$dS_t = \mu S_t dt + \sigma S_t dW_t,$$

for some constants μ and σ, where $\{W_t\}_{t\geq 0}$ is a $\mathbb{P}$-Brownian motion. Notice that this corresponds to taking $\nu = \mu - \frac{1}{2}\sigma^2$ in our calculations of §2.6. We call $\mathbb{P}$ the *market measure*. Exactly as in the discrete world, the market measure tells us which market events have positive probability, but we shall reweight the probabilities for the purposes of pricing and hedging.

Self-financing strategies

As in the discrete world, our starting point is that the market does not admit any arbitrage opportunities. Our strategy also parallels the work of Chapter 2: to obtain the time zero price of a claim, C_T, against us at time T, we seek a self-financing portfolio whose value at time T is exactly C_T. In the absence of arbitrage, the value of the claim must be the same as the cost of constructing the replicating portfolio. Of course for this argument to work, the trading strategy for this portfolio must be *previsible*. Moreover, because we are now allowed to rebalance the portfolio as often as we like, rather than just at the 'ticks' of a clock, to avoid obvious arbitrage opportunities we must introduce a further restriction on admissible trading strategies for our portfolio. We illustrate with an example.

Example 5.1.1 (The doubling strategy) *Consider the following strategy for betting on successive (independent) flips of a coin that comes up heads with probability $p > 0$. We bet $\$K$ that the first flip comes up heads. If it does come up heads then we stop, having won $\$K$. If it does not come up heads, then we bet $\$2K$ that the second flip comes up heads. If it does, then our net gain is $\$K$ and we stop. Otherwise we have lost $\$3K$ and we bet $\$4K$ that the next flip is a head. And so on. If the first $n-1$ flips all come up tails, then we have lost $\$\sum_{j=0}^{n-1} 2^j K = \$(2^n - 1)K$ and we bet $\$2^n K$ on the nth flip. Since with probability one the coin will eventually come up heads, we are guaranteed to win $\$K$. Of course, this relies on our having infinite credit. If we only have limited funds, then the apparent arbitrage opportunity disappears.*

With this example in mind, we make the following definition.

Definition 5.1.2 *A self-financing strategy is defined by a pair of predictable processes $\{\psi_t\}_{0\leq t\leq T}$, $\{\phi_t\}_{0\leq t\leq T}$, denoting the quantities of riskless and risky asset respectively held in the portfolio at time t, satisfying*

$$\int_0^T |\psi_t|\, dt + \int_0^T |\phi_t|^2\, dt < \infty$$

(with probability one), and

2

$$\psi_t B_t + \phi_t S_t = \psi_0 B_0 + \phi_0 S_0 + \int_0^t \psi_u dB_u + \int_0^t \phi_u dS_u$$

(with probability one) for all $t \in [0, T]$.

Remarks: Condition 1 ensures that the integrals in condition 2 make sense. More-over, $\int_0^t \phi_u dW_u$ will be a $\mathbb{P}$-martingale.

In differential form, condition 2 says that the value, $V_t(\psi, \phi) = \psi_t B_t + \phi_t S_t$, of the portfolio at time t satisfies

$$dV_t(\psi, \phi) = \psi_t dB_t + \phi_t dS_t,$$

that is, changes of value of the portfolio over an infinitesimal time interval are due entirely to changes in value of the assets and not to injection (or removal) of wealth from outside. □

As in the discrete setting, the key will be to work with a probability measure, $\mathbb{Q}$, equivalent to the 'market measure' $\mathbb{P}$ and under which the discounted stock price, $\{\tilde{S}_t\}_{t\geq 0}$, is a $\mathbb{Q}$-martingale. This means that it is convenient to think of the *discounted asset price process* as the object of central interest. With this in mind we prove the following continuous analogue of equation (2.5).

Lemma 5.1.3   *Let $\{\psi_t\}_{0\leq t\leq T}$ and $\{\phi_t\}_{t\geq 0}$ be predictable processes satisfying*

$$\int_0^T |\psi_t| \, dt + \int_0^T |\phi_t|^2 \, dt < \infty$$

(with probability one). Set

$$V_t(\psi, \phi) = \psi_t B_t + \phi_t S_t, \qquad \tilde{V}_t(\psi, \phi) = e^{-rt} V_t(\psi, \phi).$$

Then $\{\psi_t, \phi_t\}_{0\leq t\leq T}$ defines a self-financing strategy if and only if

$$\tilde{V}_t(\psi, \phi) = \tilde{V}_0(\psi, \phi) + \int_0^t \phi_u d\tilde{S}_u$$

with probability one for all $t \in [0, T]$.

Proof: Suppose first that the portfolio $\{\psi_t, \phi_t\}_{0\leq t\leq T}$ is self-financing. Then

$$\begin{aligned} d\tilde{V}_t(\psi, \phi) &= -re^{-rt} V_t(\psi, \phi)dt + e^{-rt} dV_t(\psi, \phi) \\ &= -re^{-rt}\left(\psi_t e^{rt} + \phi_t S_t\right) dt + e^{-rt}\psi_t d(e^{rt}) + e^{-rt}\phi_t dS_t \\ &= \phi_t\left(-re^{-rt} S_t dt + e^{-rt} dS_t\right) \\ &= \phi_t d\tilde{S}_t \end{aligned}$$

as required.

The other direction is similar and is left as an exercise. □

A strategy for
pricing

Before going further, we outline our strategy. We write C_T for the claim at time T that we are trying to replicate. It may depend on $\{S_t\}_{0 \le t \le T}$ in more complex ways than just through S_T. Suppose that *somehow* we can find a predictable process $\{\phi_t\}_{0 \le t \le T}$ such that the claim C_T, discounted, satisfies

$$\tilde{C}_T \triangleq e^{-rT}C_T = \phi_0 + \int_0^T \phi_u d\tilde{S}_u.$$

Then we can replicate the claim by a portfolio in which we hold ϕ_t units of stock and ψ_t units of cash bond at time t, where ψ_t is chosen so that

$$\tilde{V}_t(\psi, \phi) = \phi_t \tilde{S}_t + \psi_t e^{-rt} = \phi_0 + \int_0^t \phi_u d\tilde{S}_u.$$

By Lemma 5.1.3, the portfolio is then self-financing, and, moreover, $V_T = C_T$. The fair price of the claim at time zero is then $V_0 = \phi_0$.

This is fine if we know ϕ_0, but there is a quick and easy way to find the right price without explicitly finding the strategy $\{\psi_t, \phi_t\}_{t \ge 0}$. Suppose instead that we can find a probability measure, $\mathbb{Q}$, under which the discounted stock price is a martingale. Then, at least provided $\int_0^T \phi_u^2 du < \infty$,

$$\int_0^t \phi_u d\tilde{S}_u$$

will be a mean zero $\mathbb{Q}$-martingale and so

$$\mathbb{E}^{\mathbb{Q}}\big[\tilde{V}_T(\psi, \phi)\big] = \phi_0 + \mathbb{E}^{\mathbb{Q}}\left[\int_0^t \phi_u d\tilde{S}_u\right] = \phi_0.$$

So $\phi_0 = \mathbb{E}^{\mathbb{Q}}\big[\tilde{C}_T\big]$ is the fair price.

This then is entirely analogous to the pricing formula of Theorem 2.3.13. If there is a probability measure, $\mathbb{Q}$, equivalent to $\mathbb{P}$ and under which the discounted stock price is a martingale, then, provided a replicating portfolio exists, the fair time zero price of the claim is $\mathbb{E}^{\mathbb{Q}}[\tilde{C}_T]$, the discounted expected value of the claim under this measure.

We have assumed that the process $\{\phi_t\}_{t \ge 0}$ exists. We prove this (for this basic Black–Scholes market model) in Theorem 5.1.5 via an application of the Martingale Representation Theorem. First, if our pricing formula is to be of any use, we should find the *equivalent martingale measure $\mathbb{Q}$*.

An
equivalent
martingale
measure

Lemma 5.1.4 (A probability measure under which $\{\tilde{S}_t\}_{t \ge 0}$ is a martingale)   *There is a probability measure $\mathbb{Q}$, equivalent to $\mathbb{P}$, under which the discounted stock price $\{\tilde{S}_t\}_{t \ge 0}$ is a martingale. Moreover, the Radon–Nikodym derivative of $\mathbb{Q}$ with respect to $\mathbb{P}$ is given by*

$$L_t^{(\theta)} \triangleq \left.\frac{d\mathbb{Q}}{d\mathbb{P}}\right|_{\mathcal{F}_t} = \exp\left(-\theta W_t - \frac{1}{2}\theta^2 t\right),$$

where $\theta = (\mu - r)/\sigma$.

Proof: Recall that

$$dS_t = \mu S_t dt + \sigma S_t d B_t,$$

so

$$d\tilde{S}_t = \tilde{S}_t \left(-r dt + \mu dt + \sigma d W_t\right).$$

Consequently, if we set $X_t = W_t + (\mu - r)t/\sigma$,

$$d\tilde{S}_t = \tilde{S}_t \sigma d X_t.$$

Now from Theorem 4.5.1, $\{X_t\}_{t\geq 0}$ is a $\mathbb{Q}$-Brownian motion and so $\{\tilde{S}_t\}_{t\geq 0}$ is a $\mathbb{Q}$-martingale. Moreover,

$$\tilde{S}_t = \tilde{S}_0 \exp\left(\sigma X_t - \sigma^2 t/2\right).$$

□

The Fundamental Theorem of Asset Pricing

We can now prove the Fundamental Theorem of Asset Pricing in the Black–Scholes framework.

Theorem 5.1.5   *Let $\mathbb{Q}$ be the measure given by Lemma 5.1.4. Suppose that a claim at time T is given by the non-negative random variable $C_T \in \mathcal{F}_T$. If*

$$\mathbb{E}^{\mathbb{Q}}[C_T^2] < \infty,$$

then the claim is replicable and the value at time t of any replicating portfolio is given by

$$V_t = \mathbb{E}^{\mathbb{Q}}\left[e^{-r(T-t)}C_T \,\Big|\, \mathcal{F}_t\right].$$

In particular, the fair price at time zero for the option is

$$V_0 = \mathbb{E}^{\mathbb{Q}}[e^{-rT}C_T] = \mathbb{E}^{\mathbb{Q}}[\tilde{C}_T].$$

Proof: In the argument that followed Lemma 5.1.3 we showed that if we could find a process $\{\phi_t\}_{0\leq t\leq T}$ such that

$$\tilde{C}_T = \phi_0 + \int_0^T \phi_u d\tilde{S}_u,$$

then we could construct a replicating portfolio whose value at time t satisfies

$$\tilde{V}_t(\psi, \phi) = \phi_0 + \int_0^t \phi_u d\tilde{S}_u, \tag{5.1}$$

which, by the martingale property of the stochastic integral is precisely

$$\tilde{V}_t(\psi, \phi) = \mathbb{E}^{\mathbb{Q}}\left[\phi_0 + \int_0^T \phi_u d\tilde{S}_u \,\Big|\, \mathcal{F}_t\right] = \mathbb{E}^{\mathbb{Q}}\left[\tilde{C}_T \,\Big|\, \mathcal{F}_t\right] = \mathbb{E}^{\mathbb{Q}}\left[e^{-rT}C_T \,\Big|\, \mathcal{F}_t\right].$$

Undoing the discounting on $[0, t]$ gives

$$V_t(\psi, \phi) = \mathbb{E}^{\mathbb{Q}}\left[e^{-r(T-t)} C_T \,\middle|\, \mathcal{F}_t \right].$$

Let's just reassure ourselves that such a price is unique. Evidently any other replicating portfolio, $\{\hat{\psi}_t, \hat{\phi}_t\}_{0\le t\le T}$, has $V_T(\hat{\psi}, \hat{\phi}) = C_T$ and if it is self-financing (by Lemma 5.1.3) it satisfies an equation of the form (5.1). Repeating the argument above we see that we obtain the same value for any self-financing replicating portfolio.

The proof of the theorem will be complete if we can show that there *is* a predictable process $\{\phi_t\}_{0\le t\le T}$ such that

$$\tilde{C}_T = \phi_0 + \int_0^T \phi_u d\tilde{S}_u.$$

Now, by Exercise 2,

$$M_t \triangleq \mathbb{E}^{\mathbb{Q}}\left[e^{-rT} C_T \,\middle|\, \mathcal{F}_t \right]$$

is a square-integrable $\mathbb{Q}$-martingale. The natural filtration of our original Brownian motion is the same as that for the process $\{X_t\}_{t\ge 0}$ defined in Lemma 5.1.4. That is, $\{M_t\}_{t\ge 0}$ is a square-integrable 'Brownian martingale' and by the Brownian Martingale Representation Theorem 4.6.2 there exists an $\{\mathcal{F}_t\}_{0\le t\le T}$-predictable process $\{\theta_t\}_{0\le t\le T}$ such that

$$M_t = M_0 + \int_0^t \theta_s dX_s.$$

Since $d\tilde{S}_s = \sigma \tilde{S}_s dX_s$, we set

$$\phi_t = \frac{\theta_t}{\sigma \tilde{S}_t} \quad \text{and} \quad \psi_t = M_t - \phi_t \tilde{S}_t.$$

Condition 1 of Definition 5.1.2 is easily seen to be satisfied and so the strategy corresponding to $\{\psi_t, \phi_t\}_{0\le t\le T}$ defines a self-financing replicating portfolio as required.   □

Remark: The theorem that we have just proved is *very* general. Subject to our mild boundedness condition, the claim C_T could be almost arbitrarily complex provided it depends only on the path of the stock price up to time T. The price of the claim at time zero is $\mathbb{E}^{\mathbb{Q}}\left[e^{-rT} C_T \right]$ and this can be evaluated, at least numerically, even for complex claims C_T.

We have proved that not only does there exist a fair price, but moreover, we *can hedge* the claim. Its shortcoming is that although we have asserted the existence of a hedging strategy, we have not obtained an explicit expression for it. We shall find such an expression for European options, that is claims that depend only on the stock price *at maturity*, in the next section.   □

Just as in the discrete world, we have identified a procedure for valuing and replicating a claim.

Three steps to replication:

1 Find a measure $\mathbb{Q}$ under which the discounted asset price $\{\tilde{S}_t\}_{t \geq 0}$ is a martingale.
2 Form the process $M_t = \mathbb{E}^{\mathbb{Q}}\left[e^{-rT}C_T \,\middle|\, \mathcal{F}_t\right]$.
3 Find a predictable process $\{\phi_t\}_{t \geq 0}$ such that $dM_t = \phi_t d\tilde{S}_t$.

5.2   Black–Scholes price and hedge for European options

In the case of European options, that is options whose payoff depends only on the price of the underlying at the time of maturity, both the price of the option and the hedging portfolio can be obtained explicitly.

First we evaluate the price of the claim. Our assumptions are exactly those of §5.1.

Proposition 5.2.1   *The value at time t of a European option whose payoff at maturity is $C_T = f(S_T)$ is $V_t = F(t, S_t)$, where*

$$F(t, x) \;=\; e^{-r(T-t)} \int_{-\infty}^{\infty} f\left(x \exp\left((r - \sigma^2/2)(T - t) + \sigma y \sqrt{T - t}\right)\right)$$
$$\times \frac{\exp(-y^2/2)}{\sqrt{2\pi}} dy.$$

Proof: From Theorem 5.1.5 we know that the value at time t is

$$\mathbb{E}^{\mathbb{Q}}\left[e^{-r(T-t)} f(S_T) \,\middle|\, \mathcal{F}_t\right], \tag{5.2}$$

where $\mathbb{Q}$ is the martingale measure obtained in Lemma 5.1.4. Under this measure $X_t = W_t + (\mu - r)t/\sigma$ is a Brownian motion and

$$d\tilde{S}_t = \sigma \tilde{S}_t dX_t.$$

Solving this equation,

$$\tilde{S}_T = \tilde{S}_t \exp\left(\sigma(X_T - X_t) - \frac{1}{2}\sigma^2(T - t)\right).$$

We can now substitute into (5.2) to obtain

$$V_t = \mathbb{E}^{\mathbb{Q}}\left[e^{-r(T-t)} f\left(S_t e^{r(T-t)} \exp\left(\sigma(X_T - X_t) - \frac{1}{2}\sigma^2(T - t)\right)\right) \,\middle|\, \mathcal{F}_t\right].$$

Since under $\mathbb{Q}$, conditional on $\mathcal{F}_t$, $X_T - X_t$ is a normally distributed random variable with mean zero and variance $(T - t)$, we can evaluate this as

$$
\begin{aligned}
V_t &= F(t, S_t) \\
&= \int_{-\infty}^{\infty} e^{-r(T-t)} f\left(S_t e^{r(T-t)} \exp\left(\sigma z - \frac{1}{2}\sigma^2(T-t)\right)\right) \\
&\quad \times \frac{1}{\sqrt{2\pi(T-t)}} \exp\left(-\frac{z^2}{2(T-t)}\right) dz \\
&= e^{-r(T-t)} \int_{-\infty}^{\infty} f\left(S_t \exp\left(\left(r - \frac{1}{2}\sigma^2\right)(T-t) + \sigma y\sqrt{T-t}\right)\right) \\
&\quad \times \frac{1}{\sqrt{2\pi}} \exp\left(-\frac{y^2}{2}\right) dy,
\end{aligned}
$$

as required.   □

Pricing calls and puts

For European calls and puts, the function F of Proposition 5.2.1 can be calculated explicitly.

Example 5.2.2 (European call)   *In the notation of Proposition 5.2.1, suppose that* $f(S_T) = (S_T - K)_+$. *Then, writing* $\theta = (T - t)$,

$$
F(t, x) = x\Phi(d_1) - Ke^{-r\theta}\Phi(d_2), \tag{5.3}
$$

where $\Phi(\cdot)$ *is the standard normal distribution function, given by*

$$
\Phi(y) = \int_{-\infty}^{y} \frac{1}{\sqrt{2\pi}} e^{-y^2/2} dy,
$$

$$
d_1 = \frac{\log\left(\frac{x}{K}\right) + \left(r + \frac{\sigma^2}{2}\right)\theta}{\sigma\sqrt{\theta}}
$$

and $d_2 = d_1 - \sigma\sqrt{\theta}$.

Proof: Substituting for f and x in the last line of the proof of Proposition 5.2.1 we have that

$$
F(t, x) = \mathbb{E}\left[\left(xe^{\sigma\sqrt{\theta}Z - \sigma^2\theta/2} - Ke^{-r\theta}\right)_+\right], \tag{5.4}
$$

where $Z \sim N(0, 1)$. First we establish for what range of values of Z the integrand is non-zero. Rearranging,

$$
xe^{\sigma\sqrt{\theta}Z - \sigma^2\theta/2} > Ke^{-r\theta}
$$

is equivalent to

$$
Z > \frac{\log\left(\frac{K}{x}\right) + \frac{\sigma^2}{2}\theta - r\theta}{\sigma\sqrt{\theta}}.
$$

Thus the integrand in (5.4) is non-zero if $Z + d_2 \geq 0$. Using this notation

$$
\begin{aligned}
F(t, x) &= \mathbb{E}\left[\left(x e^{\sigma \sqrt{\theta} Z - \sigma^2 \theta / 2} - K e^{-r\theta}\right) \mathbf{1}_{Z + d_2 \geq 0}\right] \\
&= \int_{-d_2}^{\infty} \left(x e^{\sigma \sqrt{\theta} y - \sigma^2 \theta / 2} - K e^{-r\theta}\right) \frac{e^{-y^2/2}}{\sqrt{2\pi}} dy \\
&= \int_{-\infty}^{d_2} \left(x e^{-\sigma \sqrt{\theta} y - \sigma^2 \theta / 2} - K e^{-r\theta}\right) \frac{e^{-y^2/2}}{\sqrt{2\pi}} dy \\
&= x \int_{-\infty}^{d_2} e^{-\sigma \sqrt{\theta} y - \sigma^2 \theta / 2} \frac{e^{-y^2/2}}{\sqrt{2\pi}} dy - K e^{-r\theta} \Phi(d_2).
\end{aligned}
$$

Substituting $z = y + \sigma \sqrt{\theta}$ in the first integral in the last line we finally obtain

$$
F(t, x) = x \Phi(d_1) - K e^{-r\theta} \Phi(d_2).
$$

$\square$

Equation (5.3) is known as the *Black–Scholes pricing formula* for a European call option. The corresponding formula for a European *put* option can be found in Exercise 3.

Remarks:

1   Bachelier actually obtained a formula that looks very like this for the price of a European call option, except that the geometric Brownian motion is replaced by Brownian motion. This, however, was a fluke. Bachelier was using expectation pricing and did not have the notion of dynamic hedging.

2   Notice that the pricing formula depends on just one unknown parameter, σ, called the *volatility* by practitioners. The same will be true of our hedging portfolio. The problem that then arises is how to estimate σ from market data. The commonest approach is to use the *implied volatility*. Some options are quoted on organised markets. The price of European call and put options is an increasing function of volatility and so we can invert the Black–Scholes formula and associate an implied volatility with each option. Unfortunately, the estimate of σ obtained in this way usually depends on strike price and time to maturity. We briefly discuss the implications of this in §7.4. $\square$

Hedging calls and puts

We now turn to the problem of *hedging* European options. That is, how should we construct a portfolio that replicates the claim against us?

The Martingale Representation Theorem tells us that since the discounted option price and the discounted stock price are martingales under the same measure, one is locally just a scaled version of the other. It is this local scaling that we should like an expression for. In our discrete world of §2.5 we found ϕ_{i+1} as the ratio of the change in value of the option to that of the stock over the $(i + 1)$st tick of the clock. It is reasonable to guess then that in the continuous world ϕ_t should be the partial derivative of the option value with respect to the stock price and this is what we now prove.

Proposition 5.2.3   *In the notation of Proposition 5.2.1, the process* $\{\phi_t\}_{0 \le t \le T}$ *that determines the stock holding in the replicating portfolio of Theorem 5.1.5 is given by*

$$\phi_t = \frac{\partial F}{\partial x}(t, x)\Big|_{x=S_t}.$$

Proof: In this notation, the result of Theorem 5.1.5 becomes

$$F(t, x) = \mathbb{E}^{\mathbb{Q}}\left[e^{-r(T-t)} f(S_T)\,\Big|\, S_t = x\right]$$

where, under $\mathbb{Q}$,

$$d\tilde{S}_t = \sigma \tilde{S}_t dX_t$$

and $\{X_t\}_{0 \le t \le T}$ is a Brownian motion. Evidently

$$dS_t = r S_t dt + \sigma S_t dX_t.$$

Combining the Feynman–Kac representation and the usual product rule of differentiation, $F(t, x)$ satisfies

$$\frac{\partial F}{\partial t}(t, x) + \frac{1}{2}\sigma^2 x^2 \frac{\partial^2 F}{\partial x^2}(t, x) - r F(t, x) + rx\frac{\partial F}{\partial x}(t, x) = 0, \qquad 0 \le t \le T.$$

This is the *Black–Scholes equation*. You are asked to verify that $F(t, x)$ satisfies this equation via a different route in Exercise 4.

Define the function $\tilde{F}(t, x) = e^{-rt} F(t, xe^{rt})$, then $\tilde{V}_t = \tilde{F}(t, \tilde{S}_t)$. Observing that

$$\frac{\partial \tilde{F}}{\partial t}(t, x) = -\frac{1}{2}\sigma^2 x^2 \frac{\partial^2 \tilde{F}}{\partial x^2}(t, x)$$

and applying Itô's formula to $\tilde{F}(u, \tilde{S}_u)$ for $0 \le u \le T$,

$$\begin{aligned}
\tilde{F}(T, \tilde{S}_T) &= F(0, S_0) + \int_0^T \sigma \tilde{S}_s \frac{\partial \tilde{F}}{\partial x}(s, \tilde{S}_s) dX_s \\
&= F(0, S_0) + \int_0^T \frac{\partial \tilde{F}}{\partial x}(s, \tilde{S}_s) d\tilde{S}_s.
\end{aligned}$$

This gives

$$\phi_t = \frac{\partial \tilde{F}}{\partial x}(t, \tilde{S}_t) = \frac{\partial F}{\partial x}(t, S_t)$$

as required.   □

Example 5.2.4 (Hedging a European call)   *Using the notation of Example 5.2.2, for a European call option we obtain*

$$\frac{\partial F}{\partial x}(t, x) = \Phi(d_1).$$

Proof: Using the same notation as in Example 5.2.2 we have

$$F(t, x) = \mathbb{E}\left[\left(x \exp\left(\sigma \sqrt{\theta} Z - \sigma^2 \theta/2\right) - K\right)_+\right],$$

where $Z \sim N(0, 1)$ and $\theta = (T - t)$. Differentiating the integrand with respect to x gives $\exp\left(\sigma \sqrt{\theta} Z - \sigma^2/2\right)$ if the integrand is strictly positive and zero otherwise. Then, again using the notation of Example 5.2.2,

$$
\begin{aligned}
\frac{\partial F}{\partial x}(t, x) &= \mathbb{E}\left[\exp\left(\sigma \sqrt{\theta} Z - \sigma^2 \theta/2\right) \mathbf{1}_{Z + d_2 \geq 0}\right] \\
&= \int_{-d_2}^{\infty} \exp\left(\sigma \sqrt{\theta} y - \sigma^2 \theta/2 - y^2/2\right) \frac{1}{\sqrt{2\pi}} dy.
\end{aligned}
$$

Substituting first $u = -y$ and then $z = u + \sigma \sqrt{\theta}$ as before this reduces to $\Phi(d_1)$. So

$$\frac{\partial F}{\partial x}(t, x) = \Phi(d_1).$$

For the European put one calculates

$$\frac{\partial F}{\partial x}(t, x) = -\Phi(-d_1).$$

□

Remark: (The Greeks) The quantity $\partial F/\partial x$ is often called the *delta* of the option by practitioners. For a portfolio π of assets and derivatives, the sensitivities of the price to the parameters of the market are determined by the *Greeks*. If we write $\pi(t, x)$ for the value of the portfolio if the asset price at time t is x, then in addition to the delta, given by $\frac{\partial \pi}{\partial x}$, we have the gamma, the theta and the vega:

$$\Gamma = \frac{\partial^2 \pi}{\partial x^2}, \quad \Theta = \frac{\partial \pi}{\partial t}, \quad \mathcal{V} = \frac{\partial \pi}{\partial \sigma}.$$

□

5.3   Foreign exchange

In this section we begin our programme of increasing the financial sophistication in our models by looking at the foreign exchange market. Holding currency is a risky business, and with this risk comes a demand for derivatives. To operate in this market we should like to be able to value claims based on the future value of one unit of currency in terms of another.

The pricing problem for an exchange rate forward was solved in Exercise 13 of Chapter 1. In contrast to the pricing problem for a forward contract based on an underlying stock that pays no dividends, which we solved in §1.2, for an exchange rate forward we needed to take into account interest rates in *both* currencies. Similarly, in valuing a European call option based on the exchange rate between

Sterling and US dollars in Example 1.6.6 we needed a model for both the Sterling and the US dollar cash bond. Our Black–Scholes model for foreign exchange markets too must incorporate cash bonds in both currencies. For definiteness, we suppose that the two currencies are US dollars and pounds Sterling.

Black–Scholes currency model: We write $\{B_t\}_{t\geq 0}$ for the dollar cash bond and $\{D_t\}_{t\geq 0}$ for its Sterling counterpart. Writing E_t for the dollar worth of one pound at time t, our model is

$$\text{Dollar bond} \quad B_t = e^{rt},$$
$$\text{Sterling bond} \quad D_t = e^{ut},$$
$$\text{Exchange rate} \quad E_t = E_0 \exp\left(vt + \sigma W_t\right),$$

where $\{W_t\}_{t\geq 0}$ is a $\mathbb{P}$-Brownian motion and r, u, v and σ are constants.

We now encounter exactly the problem that we had to overcome in the discrete world: the exchange rate is *not tradable*. We must confine ourselves to operating within a single market. Let us work first from the point of view of the dollar investor. In the dollar markets, neither the Sterling cash bond nor the exchange rate is tradable. However, the product of the two, $S_t = E_t D_t$, can be thought of as a dollar tradable. The dollar investor can hold Sterling cash bonds and their dollar value is precisely S_t at time t. Moreover, any claim based on E_T can be thought of as a claim based on S_T.

We now have a set-up that precisely mirrors the basic Black–Scholes model of §5.1. From the point of view of the dollar trader there are really two processes, the dollar cash bond, $\{B_t\}_{t\geq 0}$, and the dollar value of the Sterling cash bond, $\{S_t\}_{t\geq 0}$. We can now apply the Black–Scholes methodology in this setting. Let C_T denote the claim value (in dollars) at time T.

Three steps to replication (foreign exchange):
1 Find a measure $\mathbb{Q}$ under which the (dollar bond) discounted process $\{\tilde{S}_t = B_t^{-1} S_t\}_{t\geq 0}$ is a martingale.
2 Form the process $M_t = \mathbb{E}^{\mathbb{Q}}\left[e^{-rT} C_T \mid \mathcal{F}_t\right]$.
3 Find an adapted process $\{\phi_t\}_{0\leq t\leq T}$ such that $dM_t = \phi_t d\tilde{S}_t$.

Since $S_t = E_t D_t = \exp\left((v + u)t + \sigma W_t\right)$, the process $\{S_t\}_{t\geq 0}$ is just a geometric Brownian motion and so our work of §5.1 ensures that we can indeed follow these steps.

First we apply Itô's formula to obtain the stochastic differential equation satisfied by $\{S_t\}_{t\geq 0}$:

$$dS_t = \left(v + u + \frac{1}{2}\sigma^2\right)S_t dt + \sigma S_t dW_t.$$

Now applying Lemma 5.1.4, we find that the Radon–Nikodym derivative with respect to $\mathbb{P}$ of the measure $\mathbb{Q}$ under which the dollar-discounted process $\{\tilde{S}_t\}_{t\geq 0}$ is a martingale is

$$\left.\frac{d\mathbb{Q}}{d\mathbb{P}}\right|_{\mathcal{F}_t} = L_t^{(\theta)} \triangleq \exp\left(-\theta W_t - \frac{1}{2}\theta^2 t\right),$$

where $\theta = \left(v + u + \frac{1}{2}\sigma^2 - r\right)/\sigma$. Moreover,

$$X_t \triangleq W_t + \frac{\left(v + u + \frac{1}{2}\sigma^2 - r\right)}{\sigma}t$$

is a $\mathbb{Q}$-Brownian motion.

We follow the rest of the procedure in a special case (see also Exercise 11).

Example 5.3.1 (Forward contract)   *At what price should we agree to trade Sterling at a future date T?*

Solution: Of course we already solved this problem in Exercise 13 of Chapter 1, but now rather than guessing the hedging portfolio, we follow our three steps to replication.

We have already found the measure $\mathbb{Q}$. Now if we agree to buy a unit of Sterling for K dollars at time T, then the payoff of the contract will be

$$C_T = E_T - K.$$

The value of the contract at time t is then

$$\begin{aligned} V_t &= \mathbb{E}^{\mathbb{Q}}\left[e^{-r(T-t)}C_T \,\middle|\, \mathcal{F}_t\right] \\ &= \mathbb{E}^{\mathbb{Q}}\left[e^{-r(T-t)}(E_T - K) \,\middle|\, \mathcal{F}_t\right]. \end{aligned}$$

A forward contract costs nothing at time zero and so we must choose K so that $V_0 = 0$. In other words, $K = \mathbb{E}^{\mathbb{Q}}[E_T]$. Expressing E_T as a function of X_T gives

$$E_T = E_0 \exp\left(\sigma X_T - \frac{1}{2}\sigma^2 T + (r - u)T\right),$$

and so using that $\{X_t\}_{t\geq 0}$ is a $\mathbb{Q}$-Brownian motion gives the fair value of K as

$$K = \mathbb{E}^{\mathbb{Q}}[E_T] = e^{(r-u)T}E_0.$$

Finally we find the hedging portfolio. With this choice of strike price, the value of the contract at time t is

$$V_t = \mathbb{E}\left[e^{-r(T-t)}\left(E_T - E_0 e^{(r-u)T}\right) \,\middle|\, \mathcal{F}_t\right].$$

Since $E_T = D_T^{-1} B_T \tilde{S}_T$, under $\mathbb{Q}$ we have

$$\mathbb{E}\left[E_T \mid \mathcal{F}_t\right] = e^{(r-u)(T-t)} D_t^{-1} B_t \mathbb{E}\left[\tilde{S}_T \mid \mathcal{F}_t\right] = e^{(r-u)(T-t)} D_t^{-1} B_t \tilde{S}_t = e^{(r-u)(T-t)} E_t,$$

so, substituting,

$$V_t = e^{-u(T-t)} E_t - e^{rt-uT} E_0 = e^{-uT}\left(e^{ut} E_t - e^{rt} E_0\right).$$

The dollar-discounted portfolio value is

$$M_t = e^{-rt} V_t = e^{-uT} e^{-(r-u)t} E_t - e^{-uT} E_0 = e^{-uT} \tilde{S}_t - e^{-uT} E_0.$$

The required hedging portfolio is constant, consisting of $\phi_t = e^{-uT}$ Sterling bonds and $\psi_t = -e^{-uT} E_0$ (dollar) cash bonds. $\square$

The Sterling investor

We now turn to the Sterling investor. From her point of view, tradables are quoted in pounds Sterling. Once again, in effect there are two Sterling tradables. The first is the Sterling cash bond. The second is the Sterling value of the dollar bond given by $Z_t = E_t^{-1} B_t$.

Once again we can follow our three-step replication programme. The Sterling-discounted value of the dollar bond is

$$\tilde{Z}_t = D_t^{-1} E_t^{-1} B_t = E_0^{-1} \exp\left(-\sigma W_t - (v + u - r) t\right).$$

We use Lemma 5.1.4 to see that under the measure $\mathbb{Q}^{\pounds}$ given by

$$\left.\frac{d\mathbb{Q}^{\pounds}}{d\mathbb{P}}\right|_{\mathcal{F}_t} = L_t^{(\lambda)} \triangleq \exp\left(-\lambda W_t - \frac{1}{2}\lambda^2 t\right),$$

with $\lambda = \left(v + u - r - \sigma^2/2\right)/\sigma$, $\{\tilde{Z}_t\}_{t \geq 0}$ is a martingale and

$$X_t' = W_t + \frac{\left(v + u - r - \sigma^2/2\right)}{\sigma} t$$

is a $\mathbb{Q}^{\pounds}$-Brownian motion. For the Sterling investor then the option price is

$$U_t = D_t \mathbb{E}^{\mathbb{Q}^{\pounds}}\left[D_T^{-1} E_T^{-1} C_T \mid \mathcal{F}_t\right].$$

Change of numeraire

We now have the same worry that we encountered in Exercise 15 of Chapter 1. The risk-neutral measures $\mathbb{Q}$ and $\mathbb{Q}^{\pounds}$ can be thought of as defining a probability measure on the paths followed by $\{E_t\}_{t \geq 0}$ – the only truly random part of our model – and the two measures are *different*. So do they give the same price?

To put our minds at rest we find the dollar worth of the Sterling investor's valuation, that is

$$E_t U_t = E_t D_t \mathbb{E}^{\mathbb{Q}^{\pounds}}\left[D_T^{-1} E_T^{-1} C_T \mid \mathcal{F}_t\right].$$

To compare this with our expression for V_t we express the expectation as a $\mathbb{Q}$-expectation, again using Girsanov's Theorem. Now

$$X_t' = W_t + \frac{\left(v + u - r - \frac{1}{2}\sigma^2\right)}{\sigma} t = X_t - \sigma t$$

and so

$$\zeta_t \triangleq \left.\frac{d\mathbb{Q}^{\pounds}}{d\mathbb{Q}}\right|_{\mathcal{F}_t} = \exp\left(\sigma X_t - \frac{1}{2}\sigma^2 t\right).$$

Now

$$E_t = E_0 \exp\left(\sigma X_t - \frac{1}{2}\sigma^2 t + (r - u)t\right),$$

so we can write $\zeta_t = B_t^{-1} D_t E_t$. Substituting gives

$$
\begin{aligned}
E_t U_t &= E_t D_t \mathbb{E}^{\mathbb{Q}^{\pounds}}\left[D_T^{-1} E_T^{-1} C_T \,\middle|\, \mathcal{F}_t\right] \\
&= E_t D_t \zeta_t^{-1} \mathbb{E}^{\mathbb{Q}}\left[D_T^{-1} E_T^{-1} \zeta_T C_T \,\middle|\, \mathcal{F}_t\right] \\
&= B_t \mathbb{E}^{\mathbb{Q}}\left[B_T^{-1} C_T \,\middle|\, \mathcal{F}_t\right].
\end{aligned}
$$

In other words the dollar value obtained by the Sterling investor is precisely V_t. The difference in the measures is merely an artefact of the different choice of 'reference asset' or *numeraire*.

5.4   Dividends

Our assumption so far has been that there is no value in simply holding a stock. We should now like to relax that assumption to allow the pricing and hedging of options based on equities – stocks that make periodic cash payments.

Continuous payments

It is simplest to begin with a dividend that is paid continuously. Assume as before that the stock price follows a geometric Brownian motion given by

$$S_t = S_0 \exp\left(\nu t + \sigma W_t\right),$$

but now in the infinitesimal time interval $[t, t + dt)$ the holder of the stock receives a dividend payment of $\delta S_t dt$ where δ is a constant. As always, we also assume that the market contains a riskless cash bond, $\{B_t\}_{t \geq 0}$, and we denote the continuously compounded interest rate by r.

The difficulty that we now face is that $\{S_t\}_{t \geq 0}$ does not represent the true worth of the asset: if we buy stock for price S_0 at time zero, when we sell it at time t, the value of having held it is not just $S_t - S_0$ but also the total accumulated dividends. In this model these depend on all the values that are taken by the asset price in the time interval $[0, t]$. In this sense, $\{S_t\}_{t \geq 0}$ is not *tradable*.

Our solution, just as for foreign exchange, is to translate the process into something that *is* tradable. The simplest solution is as follows. Suppose that whenever a cash dividend is paid, we immediately reinvest it in stock. The infinitesimal payout $\delta S_t dt$ will buy δdt units of stock. At time t, rather than holding one unit of stock, we hold $e^{\delta t}$ units with total worth

$$Z_t = S_0 \exp\left((\nu + \delta)t + \sigma W_t\right).$$

We regard the simple portfolio obtained by holding stock and continuously reinvesting the dividends in this way as a single asset with value Z_t at time t. There is no cost in holding this asset and it makes no dividend payments. We are back in familiar Black–Scholes country.

Remark: Because the dividend payments were a constant proportion of the stock price, it was natural to reinvest them in stock. Had payments been fixed amounts of cash, it would have been more natural to construct our 'tradable' asset as a portfolio in which dividends were immediately reinvested in bonds. This will be the situation of §5.5.   □

Any portfolio consisting of $\phi_t e^{\delta t}$ units of our original dividend-paying stock and ψ_t cash bonds at time t can be thought of as a portfolio of ϕ_t units of our new tradable asset and ψ_t units of cash bond.

We can now follow our familiar procedure.

Three steps to replication (continuous dividends): Let $\tilde{Z}_t = B_t^{-1} Z_t = e^{-rt} Z_t$.

1 Find a probability measure $\mathbb{Q}$ under which $\{\tilde{Z}_t\}_{t \geq 0}$ (with its natural filtration) is a martingale.

2 Form the discounted value process,

$$\tilde{V}_t = \mathbb{E}^{\mathbb{Q}}\!\left[e^{-rT} C_T \,\middle|\, \mathcal{F}_t\right].$$

3 Find a predictable process $\{\phi_t\}_{0 \leq t \leq T}$ such that

$$d\tilde{V}_t = \phi_t \, d\tilde{Z}_t.$$

Notice that a portfolio consisting of ϕ_t units of our tradable asset and ψ_t units of cash bond at each time $t \in [0, t]$ is *self-financing* if the value, $\{V_t\}_{t \geq 0}$, satisfies

$$dV_t = \phi_t \, dZ_t + \psi_t \, dB_t = \phi_t \, dS_t + \phi_t \delta S_t \, dt + \psi_t \, dB_t.$$

The change in value over $[t, t + dt)$ is due not only to profits and losses of trading but also to dividend payments.

Example 5.4.1 (Call option) *Suppose that a call option with strike price K and maturity T is written on the dividend-paying stock described above. What is the value of the option at time zero and what is the replicating portfolio?*

Solution: We follow our three steps to replication. First we must find the martingale measure $\mathbb{Q}$. The stochastic differential equation satisfied by $\{\tilde{Z}_t\}_{t \geq 0}$ is

$$d\tilde{Z}_t = \left(v + \delta + \frac{1}{2}\sigma^2 - r\right)\tilde{Z}_t \, dt + \sigma \tilde{Z}_t \, dW_t.$$

As usual we apply the Girsanov Theorem. Under the measure $\mathbb{Q}$ defined by

$$\left.\frac{d\mathbb{Q}}{d\mathbb{P}}\right|_{\mathcal{F}_t} = \exp\left(-\lambda W_t - \frac{1}{2}\lambda^2 t\right)$$

with $\lambda = \left(v + \delta + \frac{1}{2}\sigma^2 - r\right)/\sigma$

$$X_t = W_t + \frac{\left(v + \delta + \frac{1}{2}\sigma^2 - r\right)}{\sigma}t$$

is a Brownian motion and hence $\{\tilde{Z}_t\}_{t\geq 0}$ is a martingale. We can now just read off the price and the hedge from the corresponding formulae in Example 5.2.2 and Example 5.2.4. The value of the portfolio at time t is

$$\begin{aligned}
V_t &= e^{-r(T-t)}\mathbb{E}^{\mathbb{Q}}\left[(S_T - K)_+ \mid \mathcal{F}_t\right] \\
&= e^{-r(T-t)}\mathbb{E}^{\mathbb{Q}}\left[\left(e^{-\delta T}Z_T - K\right)_+ \mid \mathcal{F}_t\right] \\
&= e^{-r(T-t)}e^{-\delta T}\mathbb{E}^{\mathbb{Q}}\left[\left(Z_T - Ke^{\delta T}\right)_+ \mid \mathcal{F}_t\right].
\end{aligned}$$

That is the value is that of $e^{-\delta T}$ copies of a call option on $\{Z_t\}_{t\geq 0}$ with maturity T and strike $Ke^{\delta t}$. We write $F_t = e^{(r-\delta)(T-t)}S_t$ for the forward price of our underlying stock at time t (see Exercise 14). Substituting from Example 5.2.2 we obtain

$$\begin{aligned}
V_t &= e^{-\delta T}\left\{ Z_t \Phi\left(\frac{\log\left(\frac{Z_t}{Ke^{\delta T}}\right) + \left(r + \frac{\sigma^2}{2}\right)(T-t)}{\sigma\sqrt{T-t}}\right)\right.\\
&\qquad \left. - Ke^{\delta T}e^{-r(T-t)}\Phi\left(\frac{\log\left(\frac{Z_t}{Ke^{\delta T}}\right) + \left(r - \frac{\sigma^2}{2}\right)(T-t)}{\sigma\sqrt{T-t}}\right)\right\} \\
&= e^{-r(T-t)}\left\{ F_t \Phi\left(\frac{\log\left(\frac{F_t}{K}\right) + \frac{1}{2}\sigma^2(T-t)}{\sigma\sqrt{T-t}}\right)\right.\\
&\qquad \left. - K\Phi\left(\frac{\log\left(\frac{F_t}{K}\right) - \frac{1}{2}\sigma^2(T-t)}{\sigma\sqrt{T-t}}\right)\right\}.
\end{aligned}$$

Now using Example 5.2.4 we see that the replicating portfolio should consist of $e^{-\delta T}\phi_t$ units of our tradable asset Z_t at time t where

$$\phi_t = \Phi\left(\frac{\log\left(\frac{F_t}{K}\right) + \frac{1}{2}\sigma^2(T-t)}{\sigma\sqrt{T-t}}\right),$$

corresponding to

$$e^{-\delta(T-t)}\Phi\left(\frac{\log\left(\frac{F_t}{K}\right) + \frac{1}{2}\sigma^2(T-t)}{\sigma\sqrt{T-t}}\right)$$

units of the dividend-paying asset. The bond holding in the portfolio will be

$$\psi_t = -Ke^{-rT}\,\Phi\left(\frac{\log\left(\frac{F_t}{K}\right) - \frac{1}{2}\sigma^2(T-t)}{\sigma\sqrt{T-t}}\right).$$

□

Example 5.4.2 (Guaranteed equity profits)   *Let $\{S_t\}_{t\geq 0}$ denote the value of the UK FTSE stock index. Suppose that we buy a five-year contract that pays out Z defined to be 90% of the ratio of the terminal and initial values of the FTSE if this value is in the interval [1.3, 1.8], 1.3 if Z < 1.3 and 1.8 if Z > 1.8. What is the value of this contract at time zero?*

Solution: The claim C_T is

$$C_T = \min\left\{\max\left\{1.3, 0.9\frac{S_T}{S_0}\right\}, 1.8\right\},$$

where T is five years. Since the claim is based on a ratio, without loss of generality we set $S_0 = 1$. As FTSE is composed of one hundred different stocks, their separate dividend payments will approximate a continuously paying stream. We assume the following data:

$$
\begin{aligned}
\text{FTSE drift} \quad & \mu = 7\%, \\
\text{FTSE volatility} \quad & \sigma = 15\%, \\
\text{FTSE dividend yield} \quad & \delta = 4\%, \\
\text{UK interest rate} \quad & r = 6.5\%.
\end{aligned}
$$

We can rewrite the claim as the sum of some cash plus the difference in the payout of two FTSE calls,

$$C_T = 1.3 + 0.9\left\{(S_T - 1.444)_+ - (S_T - 2)_+\right\}.$$

Now the forward price for S_t is

$$F_t = e^{(r-\delta)(T-t)}S_0 = 1.133,$$

and so using the call price formula for continuous dividend-paying stocks of Example 5.4.1 we can value these calls at 0.0422 and 0.0067 (per unit) at time zero. The value of our contract at time zero is then

$$1.3e^{-rT} + 0.9\,(0.0422 - 0.0067) = 0.9712.$$

□

Periodic dividends   In practice, an individual stock does not pay dividends continuously, but rather at regular intervals. Suppose that the times of the payments are known in advance to be $T_1, T_2, \ldots$ and that at each time T_i the current holder of the equity receives a payment of δS_{T_i}. As shown in Exercise 7 of Chapter 2, in the absence of arbitrage, the stock price must instantaneously decrease by the same amount. So at any of the times T_i, the dividend payout is exactly equal to the instantaneous decrease in the stock price. Between payouts we assume our usual geometric Brownian motion model.

Equity model with periodic dividends:　At deterministic times $T_1, T_2, \ldots$ the equity pays a dividend of a fraction δ of the stock price which was current just before the dividend is paid. The stock price itself is modelled as

$$S_t = S_0 (1 - \delta)^{n[t]} \exp (\nu t + \sigma W_t),$$

where $n[t] = \max\{i : T_i \leq t\}$ is the number of dividend payments made by time t. There is also a riskless cash bond $B_t = \exp (rt)$.

At first sight it looks as though we have two problems. First, although between the times T_i our stock price follows the usual geometric Brownian motion model, at those times it has discontinuous jumps. This doesn't fit our framework. Secondly, as for continuous dividends, the stock price process $\{S_t\}_{t \geq 0}$ does not reflect the true value of the stock. However, by adapting our strategy for the continuous dividends case and reinvesting all dividend payments in stock, we'll overcome both of these obstacles.

We define $\{Z_t\}_{t \geq 0}$ to be the value of the portfolio that starts with one unit of stock at time zero and every time the stockholding pays a dividend it is reinvested by buying more stock. The first dividend payment is δS_{T_1-}, δ times the stock price immediately prior to the payment. Immediately after the payment of the first dividend, the stock price jumps to $S_{T_1+} = (1 - \delta) S_{T_1-}$, so our dividend payment will buy us an additional $\delta/(1 - \delta)$ units of stock, thereby increasing our total stock holding by a factor of $1/(1 - \delta)$. At time t the portfolio will therefore consist of $1/(1 - \delta)^{n[t]}$ units of stock. Thus

$$Z_t = (1 - \delta)^{-n[t]} S_t = S_0 \exp (\nu t + \sigma W_t).$$

As before we think of our portfolio $\{Z_t\}_{t \geq 0}$ as a non-dividend-paying asset and so our market consists of two tradable assets, the portfolio $\{Z_t\}_{t \geq 0}$ and a riskless cash bond $\{B_t\}_{t \geq 0}$, and we are back in familiar territory.

We mimic exactly what we did for continuous dividend payments. A portfolio consisting of ϕ_t units of Z_t and ψ_t in cash bonds at time t is equivalent to $(1 - \delta)^{-n[t]} \phi_t$ units of the dividend-paying underlying stock, S_t, and ψ_t units of cash bond.

The measure $\mathbb{Q}$ that makes the discounted process $\{\tilde{Z}_t\}_{t \geq 0}$ a martingale satisfies

$$\left. \frac{d\mathbb{Q}}{d\mathbb{P}} \right|_{\mathcal{F}_t} = \exp \left(-\lambda W_t - \frac{1}{2} \lambda^2 t \right)$$

with $\lambda = \left(v + \tfrac{1}{2}\sigma^2 - r\right)/\sigma$. Rewriting the claim as a function of Z_T we can use the classical Black–Scholes analysis to price and hedge the option.

Example 5.4.3 (Forward price)   *Find the fair price to be written into a forward contract on a stock that pays periodic dividends.*

Solution: The value of the contract at time T is $C_T = S_T - K$. We seek K so that the time zero value of the contract is zero. As usual, the value at time t is the discounted expected value of the claim under the martingale measure $\mathbb{Q}$. That is

$$
\begin{aligned}
V_t &= \mathbb{E}^{\mathbb{Q}}\left[e^{-r(T-t)}\left(S_T - K\right)\middle|\mathcal{F}_t\right] \\
&= \mathbb{E}^{\mathbb{Q}}\left[e^{-r(T-t)}\left((1-\delta)^{n[T]}Z_T - K\right)\middle|\mathcal{F}_t\right] \\
&= (1-\delta)^{n[T]}Z_t - Ke^{-r(T-t)} \\
&= (1-\delta)^{n[T]-n[t]}S_t - Ke^{-r(T-t)}.
\end{aligned}
$$

The K for which this is zero at time zero is

$$
K = e^{rT}(1-\delta)^{n[T]}S_0. \tag{5.5}
$$

$\square$

5.5 Bonds

A pure discount bond is a security that pays off one unit at some future time T. Most market bonds also pay off a series of smaller amounts, c, at predetermined times $T_1, T_2, \ldots, T_n$. Such *coupon* payments resemble dividend payments except that the amount of the coupon is known in advance.

So far we have considered only a riskless cash bond in which the interest rate too is known in advance. In real markets, uncertainty in interest rates causes the price of bonds to move randomly as well. In order to keep the book to a reasonable length we do not intend to enter into a full account of bond market models. An excellent introduction can be found in Baxter & Rennie (1996). So for the purposes of this section, we are going to take a schizophrenic attitude to interest rates. We'll assume that we have a riskless cash bond following $B_t = e^{rt}$, but also a stochastically varying coupon bond whose price between coupon payments evolves as a geometric Brownian motion. Clearly there are links between the short term interest rate and bond prices, but over short time horizons, many practitioners ignore them. In effect we are thinking of a coupon bond as an asset paying predetermined cash dividends at times $T_1, T_2, \ldots, T_n$ where we assume $T_n < T$. Writing $I(t) = \min\{i : t < T_i\}$, the bond price satisfies

$$
S_t = \sum_{i=I(t)}^{n} ce^{-r(T_i - t)} + A\exp\left(vt + \sigma W_t\right),
$$

for some constants A, v and σ.

As for dividend-paying stock, the price process $\{S_t\}_{t\geq 0}$ is discontinuous at the coupon payment times. Once again however we can manufacture a continuous non-dividend-paying asset from $\{S_t\}_{t\geq 0}$. Whereas when our dividend payment was a fraction of the stock price it was natural to reinvest it in stock, now, since our coupons are fixed cash payments, we invest them in the riskless cash bond. With this investment strategy the coupon paid at time T_i then has value $ce^{-r(T_i-t)}$ for *all* $t \in [0, T]$ and so the portfolio constructed in this way has value

$$Z_t = \sum_{i=1}^{n} ce^{-r(T_i-t)} + A \exp\left(\nu t + \sigma W_t\right).$$

Exactly as before we think of our market as consisting of the riskless cash bond $\{B_t\}_{t\geq 0}$ and the tradable asset $\{Z_t\}_{t\geq 0}$.

As usual we want to find $\mathbb{Q}$ under which the discounted asset price $\{\tilde{Z}_t\}_{t\geq 0}$ is a martingale. But $\tilde{Z}_t$ is just the constant cash sum $\sum_{i=1}^{n} ce^{-rT_i}$ plus the geometric Brownian motion $A \exp\left((\nu - r)t + \sigma W_t\right)$. This will be a $\mathbb{Q}$-martingale if

$$\left.\frac{d\mathbb{Q}}{d\mathbb{P}}\right|_{\mathcal{F}_t} = \exp\left(-\lambda W_t - \frac{1}{2}\lambda^2 t\right)$$

where $\lambda = \left(\nu + \frac{1}{2}\sigma^2 - r\right)/\sigma$. Under $\mathbb{Q}$,

$$X_t = W_t + \frac{\left(\nu + \frac{1}{2}\sigma^2 - r\right)}{\sigma}t$$

is a Brownian motion. The value at time t of an option with payoff C_T is now

$$\mathbb{E}^{\mathbb{Q}}\left[e^{-r(T-t)}C_T \,\middle|\, \mathcal{F}_t\right].$$

Under $\mathbb{Q}$, the price of the bond at time T is just

$$S_T = A \exp\left(\left(r - \frac{1}{2}\sigma^2\right)T + \sigma X_T\right).$$

The forward price for S_T at time zero is $F = Ae^{rT}$ and the value of a call on S_T with strike price K at maturity T is

$$e^{-rT}\left\{F\Phi\left(\frac{\log\left(\frac{F}{K}\right) + \frac{1}{2}\sigma^2 T}{\sigma\sqrt{T}}\right) - K\Phi\left(\frac{\log\left(\frac{F}{K}\right) - \frac{1}{2}\sigma^2 T}{\sigma\sqrt{T}}\right)\right\};$$

see Exercise 20.

5.6 Market price of risk

A definite pattern has emerged. Given a non-tradable stock, we have tied it to a portfolio that can be thought of as a tradable, found the martingale measure corresponding to that tradable process and used that measure to price the option. In deciding what is tradable and what is not we have used only common sense. Indeed it is not something that can be reduced purely to mathematics, but if we decide that an asset with price $\{S_t\}_{t\geq 0}$ is truly tradable and we have a riskless cash bond, $\{B_t\}_{t\geq 0}$, we should like to determine the class of tradables within the market that they create.

Martingales
and
tradables

Suppose that $\tilde{S}_t = B_t^{-1} S_t$ is the discounted price of our tradable asset at time t. Let $\mathbb{Q}$ be a measure under which $\{\tilde{S}_t\}_{t\geq 0}$ is a martingale. If we take another process, $\{V_t\}_{t\geq 0}$ for which the discounted process $\{\tilde{V}_t\}_{t\geq 0}$ given by $\tilde{V}_t = B_t^{-1} V_t$ is a $\left(\mathbb{Q}, \{\mathcal{F}_t^{\tilde{S}}\}_{t\geq 0}\right)$-martingale, then is $\{V_t\}_{t\geq 0}$ tradable?

Our strategy is to construct a self-financing portfolio consisting of our tradable asset and the tradable discounting process whose value at time t is always exactly V_t. As usual, the first step is an application of the Martingale Representation Theorem. Provided that $B_t^{-1} S_t$ has non-zero volatility, we can find an $\{\mathcal{F}_t^{\tilde{S}}\}_{t\geq 0}$-previsible process $\{\phi_t\}_{t\geq 0}$ such that

$$d\tilde{V}_t = \phi_t d\tilde{S}_t. \tag{5.6}$$

Taking our cue from the construction of the portfolio replicating an option in §5.1, we create a portfolio that at time t consists of ϕ_t units of the tradable S_t and $\psi_t = \tilde{V}_t - \phi_t \tilde{S}_t$ units of (the tradable) B_t. The value of this portfolio at time t is exactly V_t. We must check that it is self-financing. Now

$$
\begin{aligned}
dV_t &= B_t d\tilde{V}_t + \tilde{V}_t dB_t && \text{(integration by parts)} \\
&= B_t \phi_t d\tilde{S}_t + \tilde{V}_t dB_t && \text{(equation (5.6))} \\
&= B_t \phi_t d\tilde{S}_t + \left(\psi_t + \phi_t \tilde{S}_t\right) dB_t \\
&= \phi_t \left(B_t d\tilde{S}_t + \tilde{S}_t dB_t\right) + \psi_t dB_t \\
&= \phi_t dS_t + \psi_t dB_t && \text{(integration by parts)}
\end{aligned}
$$

and so the change in value of the portfolio over any infinitesimal time interval is due to changes in asset prices. That is we have the self-financing property and $\{V_t\}_{t\geq 0}$ is indeed tradable.

What about the other way round? Suppose that $\{B_t^{-1} V_t\}_{t\geq 0}$ were *not* a $\left(\mathbb{Q}, \{\mathcal{F}_t^{\tilde{S}}\}_{t\geq 0}\right)$-martingale. Then there would have to be times $s < T$ such that with positive probability

$$B_s^{-1} V_s \neq \mathbb{E}^{\mathbb{Q}}\left[B_T^{-1} V_T \,\middle|\, \mathcal{F}_s^{\tilde{S}}\right].$$

Suppose that $\{V_t\}_{t\geq 0}$ were tradable. We can construct a process $\{U_t\}_{t\geq 0}$ by setting

$$U_t = B_t \mathbb{E}^{\mathbb{Q}}\left[B_T^{-1} V_T \,\middle|\, \mathcal{F}_t^{\tilde{S}}\right].$$

Since $\{B_t^{-1} U_t\}_{t\geq 0}$ is a $\left(\mathbb{Q}, \{\mathcal{F}_t^{\tilde{S}}\}_{t\geq 0}\right)$-martingale, we know that $\{U_t\}_{t\geq 0}$ is tradable. That is we have two tradables that take the same value at time T, but with positive probability take *different* values at an earlier time s. Exercise 21 shows that in the absence of arbitrage this is a contradiction. So if $\{B_t^{-1} V_t\}_{t\geq 0}$ is not a martingale, then $\{V_t\}_{t\geq 0}$ is *not* a tradable.

Of course, since interest rates are deterministic, $\mathcal{F}_t^{\tilde{S}} = \mathcal{F}_t^{S}$ and so we have proved the following lemma.

Lemma 5.6.1   *Given a riskless cash bond $\{B_t\}_{t\geq 0}$ and a tradable asset $\{S_t\}_{t\geq 0}$, a process $\{V_t\}_{t\geq 0}$ represents a tradable asset if and only if the discounted value*

$\{B_t^{-1}V_t\}_{t\geq 0}$ *is a* $\left(\mathbb{Q}, \{\mathcal{F}_t^S\}_{t\geq 0}\right)$*-martingale where* $\mathbb{Q}$ *is the measure under which the discounted asset price* $\{B_t^{-1}S_t\}_{t\geq 0}$ *is a martingale.*

Tradables and the market price of risk

Suppose that we have two tradable risky securities $\{S_t^1\}_{t\geq 0}$ and $\{S_t^2\}_{t\geq 0}$ in a single Black–Scholes market – that is they are both functions of the same Brownian motion. We define them both via their stochastic differential equations,

$$dS_t^i = \mu_i S_t^i dt + \sigma_i S_t^i dW_t.$$

In order for both to be tradable, Lemma 5.6.1 tells us that they must both be martingales with respect to the *same* measure $\mathbb{Q}$. Assuming that $B_t = e^{rt}$ we must have that

$$X_t = W_t + \left(\frac{\mu_i - r}{\sigma_i}\right)t$$

is a $\mathbb{Q}$-Brownian motion for $i = 1, 2$. This can only be the case if

$$\frac{\mu_1 - r}{\sigma_1} = \frac{\mu_2 - r}{\sigma_2}.$$

Economists attach a meaning to this quantity. If we think of μ as the rate of growth of the tradable, r as the rate of growth of the riskless bond and σ as a measure of the risk of the asset, then

$$\gamma = \frac{\mu - r}{\sigma}$$

is the excess rate of return (above the risk-free rate) per unit of risk. As such, it is often called the *market price of risk*. It is also known as the *Sharpe ratio*. In such a simple market, every tradable asset should have the same market price of risk, otherwise there would be arbitrage opportunities.

Of course, γ is precisely the change of drift in the underlying Brownian motion when we change measure from $\mathbb{P}$ (the market measure) to $\mathbb{Q}$ (the martingale measure). However, this appealing economic interpretation of γ does *not* provide a new argument for using $\mathbb{Q}$. It is *replication* that makes the Black–Scholes analysis work. Without a replicating portfolio our arbitrage arguments collapse.

Exercises

1   Suppose that an asset price S_t is such that $dS_t = \mu S_t dt + \sigma S_t dW_t$, where $\{W_t\}_{t\geq 0}$ is, as usual, standard $\mathbb{P}$-Brownian motion. Let r denote the risk-free interest rate. The price of a riskless asset then follows $dB_t = r B_t dt$. We write $\{\psi_t, \phi_t\}$ for the portfolio consisting of ψ_t units of the riskless asset B_t and ϕ_t units of S_t at time t. For each of the following choices of ϕ_t, find ψ_t so that the portfolio $\{\psi_t, \phi_t\}$ is self-financing. (Recall that the value of the portfolio at time t is $V_t = \psi_t B_t + \phi_t S_t$ and that the portfolio is self-financing if $dV_t = \psi_t dB_t + \phi_t dS_t$.)

(a) $\phi_t = 1$,
(b) $\phi_t = \int_0^t S_u du$,
(c) $\phi_t = S_t$.

2   Let $\{\mathcal{F}_t\}_{t\geq 0}$ be the natural filtration associated with a $\mathbb{P}$-Brownian motion $\{W_t\}_{t\geq 0}$. Show that if $\mathbb{Q}$ is a probability measure equivalent to $\mathbb{P}$ and H_T is an $\mathcal{F}_T$-measurable random variable with $\mathbb{E}^{\mathbb{Q}}\left[H_T^2\right] < \infty$ then

$$M_t \triangleq \mathbb{E}^{\mathbb{Q}}\left[H_T | \mathcal{F}_t\right]$$

defines a square-integrable $\mathbb{Q}$-martingale.

3   Show that, in the notation of Example 5.2.2, the Black–Scholes price at time t of a European put option with strike K and maturity T is $F(t, S_t)$ where

$$F(t, x) = Ke^{-r\theta}\Phi(-d_2) - x\Phi(-d_1).$$

4   Suppose that the value of a European call option can be expressed as $V_t = F(t, S_t)$ (as we prove in Proposition 5.2.3). Then $\tilde{V}_t = e^{-rt} V_t$, and we may define $\tilde{F}$ by

$$\tilde{V}_t = \tilde{F}(t, \tilde{S}_t).$$

Under the risk-neutral measure, the discounted asset price follows $d\tilde{S}_t = \sigma \tilde{S}_t dX_t$, where (under this probability measure) $\{X_t\}_{t\geq 0}$ is a standard Brownian motion.

(a) Find the stochastic differential equation satisfied by $\tilde{F}(t, \tilde{S}_t)$.
(b) Using the fact that $\tilde{V}_t$ is a martingale under the risk-neutral measure, find the partial differential equation satisfied by $\tilde{F}(t, x)$, and hence show that

$$\frac{\partial F}{\partial t} + \frac{1}{2}\sigma^2 x^2 \frac{\partial^2 F}{\partial x^2} + rx\frac{\partial F}{\partial x} - rF = 0.$$

This is the *Black–Scholes equation*.

5   *Delta hedging* The following derivation of the Black–Scholes equation is very popular in the finance literature. We will suppose, as usual, that an asset price follows a geometric Brownian motion. That is, there are parameters μ, σ, such that

$$dS_t = \mu S_t dt + \sigma S_t dW_t.$$

Suppose that we are trying to value a European option based on this asset. Let us denote the value of the option at time t by $V(t, S_t)$. We know that at time T, $V(T, S_T) = f(S_T)$, for some function f.

(a) Using Itô's formula express V as the solution to a stochastic differential equation.
(b) Suppose that a portfolio, whose value we denote by π, consists of one option and a (negative) quantity $-\delta$ of the asset. Assuming that the portfolio is self-financing, find the stochastic differential equation satisfied by π.
(c) Find the value of δ for which the portfolio you have constructed is 'instantaneously riskless', that is for which the stochastic term vanishes.
(d) An instantaneously riskless portfolio must have the same rate of return as the risk-free interest rate. Use this observation to find a (deterministic) partial differential equation for the $V(t, x)$. Notice that this is the Black–Scholes equation obtained in Exercise 4.

Of course δ is precisely the stock holding in our replicating portfolio. In fact this derivation is not entirely satisfactory as it can be checked that the portfolio that we have constructed is *not* self-financing, violating our assumption in (b). A rigorous approach requires a portfolio consisting of one option, $-\delta$ assets and $e^{-rt} (V(t, S_t) - \delta S_t)$ cash bonds at time t.

6   Calculate the values of the Greeks for a European call option with strike price K at maturity time T.

7   An alternative approach to solving the Black–Scholes equation is to transform it via a change of variables into the heat equation. Suppose that $F(t, x) : [0, T] \times [0, \infty) \to \mathbb{R}$ satisfies

$$\frac{\partial F}{\partial t}(t, x) + \frac{1}{2}\sigma^2 x^2 \frac{\partial^2 F}{\partial x^2}(t, x) + rx\frac{\partial F}{\partial x}(t, x) - rF(t, x) = 0,$$

subject to the boundary conditions

$$F(t, 0) = 0, \qquad \frac{F(t, x)}{x} \to 1 \text{ as } x \to \infty, \qquad F(T, x) = (x - K)_+.$$

That is F solves the Black–Scholes equation with the boundary conditions appropriate for pricing a European call option with strike price K at time T.

(a) Show that the change of variables

$$x = Ke^y, \qquad t = T - \frac{2\tau}{\sigma^2}, \qquad F = Kv(\tau, y)$$

results in the equation

$$\frac{\partial v}{\partial \tau}(\tau, y) = \frac{\partial^2 v}{\partial y^2}(\tau, y) + (k - 1)\frac{\partial v}{\partial y}(\tau, y) - kv(\tau, y), \qquad y \in \mathbb{R}, \tau \in \left[0, \frac{1}{2}\sigma^2 T\right],$$

where $k = 2r/\sigma^2$ and $v(0, y) = (e^y - 1)_+$.

(b) Now set $v(\tau, y) = e^{\alpha y + \beta \tau} u(\tau, y)$. Find α and β such that

$$\frac{\partial u}{\partial \tau}(\tau, y) = \frac{\partial^2 u}{\partial y^2}(\tau, y), \qquad y \in \mathbb{R},$$

and find the corresponding initial condition for u.

(c) Solve for u and retrace your steps to obtain the Black–Scholes pricing formula for a European call option.

8   Show that, for each constant A, $V(t, x) = Ax$ and $V(t, x) = Ae^{rt}$ are both exact solutions of the Black–Scholes differential equation. What do they represent and what is the hedging portfolio in each case?

9   Find the most general solution of the Black–Scholes equation that has the special form

(a) $V(t, x) = V(x)$,

(b) $V(t, x) = f(t)g(x)$.

These are examples of similarity solutions. The solutions in (a) give prices of *perpetual* options.

10   Let $C(t, S_t)$ and $P(t, S_t)$ denote the values of a European call and put option with the same exercise price, K, and expiry time, T. Show that $C(t, x) - P(t, x)$ also satisfies the Black–Scholes equation with the final data $C(T, x) - P(T, x) = x - K$. Deduce that $x - Ke^{-r(T-t)}$ is also a solution of the Black–Scholes equation. Interpret these results financially.

11   Assuming the model of §5.3, find the Black–Scholes price for a Sterling call option that gives us the right to buy a pound Sterling at time T for K dollars. What is the corresponding hedging portfolio?

12   Check that the Sterling and dollar investors of §5.3 use exactly the same replicating strategy.

13   Suppose that the US dollar/Japanese Yen exchange rate follows the stochastic differential equation

$$dS_t = \mu S_t dt + \sigma S_t dW_t$$

for some constants μ and σ. You are told that the expected \$/¥ and ¥/\$ exchange rates in one years time are both $2S_0$. Is this possible?

14   In our usual notation suppose that an asset price follows geometric Brownian motion with $S_t = S_0 \exp(vt + \sigma W_t)$ at time t. If in each infinitesimal time interval the asset pays to its holder a dividend of $\delta S_t dt$, find an expression for the fair price in a forward contract based on the stock with maturity time T. What is the corresponding hedging portfolio?

15   What is the 'put–call parity' relation for the market in Exercise 14?

16   Suppose that in valuing the contract in Example 5.4.2 we had failed to take account of the dividend stream from the constituent stocks of the FTSE. Find a financial argument to indicate whether the price obtained for the contract will be too high or too low. Find the exact value that we would have obtained for the contract.

17   Suppose that V_t is the value of a self-financing portfolio consisting of ϕ_t units of stock that pays periodic dividends, as in §5.4, and ψ_t units of cash bond. Find the differential equation that characterises the self-financing property of V_t in this setting.

18   Find a portfolio that replicates the forward of Example 5.4.3.

19   Value and hedge a European call option with maturity T and strike K based on the periodic dividend-paying stock of §5.4. Express your answer in terms of the usual Black–Scholes formula evaluated on the forward price of equation (5.5).

20   Check the forward price and the value of a call option claimed in §5.5 and find the corresponding self-financing replicating portfolios.

21   Show that if two tradable assets have the same value at time T, but with positive probability take different values at time $s < T$, then there are arbitrage opportunities in the market.

6 Different payoffs

Most of the concrete examples of options considered so far have been the standard examples of calls and puts. Such options have liquid markets, their prices are fairly well determined and margins are competitive. Any option that is not one of these *vanilla* calls or puts is called an exotic option. Such options are introduced to extend a bank's product range or to meet hedging and speculative needs of clients. There are usually no markets in these options and they are bought and sold purely 'over the counter'. Although the principles of pricing and hedging exotics are exactly the same as for vanillas, risk management requires care. Not only are these exotic products much less liquid than standard options, but they often have discontinuous payoffs and so can have huge 'deltas' close to the expiry time making them difficult to hedge.

This chapter is devoted to examples of exotic options. The simplest exotics to price and hedge are *packages*, that is, options for which the payoff is a combination of our standard 'vanilla' options and the underlying asset. We already encountered such options in §1.1. We relegate their valuation to the exercises. The next simplest examples are European options, meaning options whose payoff is a function of the stock price at the maturity time. The payoffs considered in §6.1 are discontinuous and we discover potential hedging problems. In §6.2 we turn our attention to multistage options. Such options allow decisions to be made or stipulate conditions at intermediate dates during their lifetime. The rest of the chapter is devoted to path-dependent options. In §6.3 we use our work of §3.3 to price lookback and barrier options. Asian options, whose payoff depends on the average of the stock price over the lifetime of the option, are discussed briefly in §6.4 and finally §6.5 is a very swift introduction to pricing American options in continuous time.

6.1 European options with discontinuous payoffs

We work in the basic Black–Scholes framework. That is, our market consists of a riskless cash bond whose value at time t is $B_t = e^{rt}$ and a single risky asset whose price, $\{S_t\}_{t \geq 0}$, follows a geometric Brownian motion.

In §5.2 we established explicit formulae for both the price and the hedging portfolio for European options within this framework. Specifically, if the payoff of the option at the maturity time T is $C_T = f(S_T)$ then for $0 \le t \le T$ the value of the option at time t is

$$
\begin{aligned}
V_t &= F(t, S_t) = \mathbb{E}^{\mathbb{Q}}\left[e^{-r(T-t)} f(S_T)\,\middle|\,\mathcal{F}_t\right] \\
&= e^{-r(T-t)} \int_{-\infty}^{\infty} f\left(S_t \exp\left(\left(r - \frac{\sigma^2}{2}\right)(T-t) + \sigma y\sqrt{T-t}\right)\right) \\
&\qquad\qquad \times \frac{1}{\sqrt{2\pi}} \exp\left(-\frac{y^2}{2}\right) dy,
\end{aligned}
\tag{6.1}
$$

where $\mathbb{Q}$ is the martingale measure, and the claim $f(S_T)$ can be replicated by a portfolio consisting at time t of ϕ_t units of stock and $\psi_t = e^{-rt}(V_t - \phi_t S_t)$ cash bonds where

$$
\phi_t = \left.\frac{\partial F}{\partial x}(t, x)\right|_{x=S_t}.
\tag{6.2}
$$

Mathematically, other than the issue of actually *evaluating* the integrals, that would appear to be the end of the story. However, as we shall see, rather more careful consideration of our assumptions might lead us to doubt the usefulness of these formulae when the payoff is a discontinuous function of S_T.

Digitals and pin risk

Example 6.1.1 (Digital options)   *The payoff of a digital option, also sometimes called a binary option or a cash-or-nothing option, is given by a Heaviside function. For example, a digital call option with strike price K at time T has payoff*

$$
C_T = \begin{cases} 1 & \text{if } S_T \ge K, \\ 0 & \text{if } S_T < K \end{cases}
$$

at maturity. Find the price and the hedge for such an option.

Solution: In order to implement the formula (6.1) we must establish the range of y for which

$$
S_t \exp\left(\left(r - \frac{\sigma^2}{2}\right)(T-t) + \sigma y\sqrt{T-t}\right) > K.
$$

Rearranging we see that this holds for $y > d$ where

$$
d = \frac{1}{\sigma\sqrt{T-t}}\left(\log\left(\frac{K}{S_t}\right) - \left(r - \frac{\sigma^2}{2}\right)(T-t)\right).
$$

Writing Φ for the normal distribution function and substituting in equation (6.1) we obtain

$$
\begin{aligned}
V_t &= e^{-r(T-t)} \int_{d}^{\infty} \frac{1}{\sqrt{2\pi}} e^{-y^2/2}\,dy = e^{-r(T-t)} \int_{-\infty}^{-d} \frac{1}{\sqrt{2\pi}} e^{-y^2/2}\,dy \\
&= e^{-r(T-t)}\Phi(-d) = e^{-r(T-t)}\Phi(d_2),
\end{aligned}
$$

where

$$d_2 = \frac{1}{\sigma\sqrt{T-t}}\left(\log\left(\frac{S_t}{K}\right) + \left(r - \frac{\sigma^2}{2}\right)(T-t)\right),$$

as in Example 5.2.2.

Now we turn to the hedge. By (6.2), the stock holding in our replicating portfolio at time t is

$$\begin{aligned}\phi_t &= e^{-r(T-t)}\frac{1}{S_t}\frac{1}{\sqrt{2\pi(T-t)}\sigma} \\ &\quad \times \exp\left(-\frac{1}{2(T-t)\sigma^2}\left(\log\left(\frac{S_t}{K}\right) + \left(r - \frac{\sigma^2}{2}\right)(T-t)\right)^2\right).\end{aligned}$$

Now as $t \uparrow T$, this converges to $1/K$ times the delta function concentrated on $S_T = K$. Consider what this means for the replicating portfolio as $t \uparrow T$. Away from $S_t = K$, ϕ_t is close to zero, but if S_t is close to K the stock holding in the portfolio will be very large. Now if near expiry the asset price is close to K, there is a high probability that its value will cross the value $S_t = K$ many times before expiry. But if the asset price oscillates around the strike price close to expiry our prescription for the hedging portfolio will tell us to rapidly buy and sell large numbers of the underlying asset. Since markets are not the perfect objects envisaged in our Black–Scholes model and we cannot instantaneously buy and sell, risk from small asset price changes (not to mention transaction costs) can easily outweigh the maximum liability that we are exposed to by having sold the digital. This is known as the *pin risk* associated with the option.   □

If we can overcome our misgivings about the validity of the Black–Scholes price for digitals, then we can use them as building blocks for other exotics. Indeed, since the option with payoff $\mathbf{1}_{[K_1,K_2]}(S_T)$ at time T can be replicated by buying a digital with strike K_2 and maturity T and selling a digital with strike K_1 and maturity T, in theory we could price any European option by replicating it by (possibly infinite) linear combinations of digitals.

6.2 Multistage options

Some options either allow decisions to be made or stipulate conditions at intermediate dates during their lifetime. An example is the forward start option of Exercise 3 of Chapter 2. To illustrate the procedure for valuation of multistage options, we find the Black–Scholes price of a forward start.

Example 6.2.1 (Forward start option)   *Recall that a forward start option is a contract in which the holder receives, at time T_0, at no extra cost, an option with expiry date $T_1 > T_0$ and strike price equal to S_{T_0}. If the risk-free rate is r find the Black–Scholes price, V_t, of such an option at times $t < T_1$.*

Solution: First suppose that $t \in [T_0, T_1]$. Then by time t we know S_{T_0} and so the value of the option is just that of a European call option with strike S_{T_0} and maturity T_1, namely

$$V_t = e^{-r(T_1-t)} \mathbb{E}^Q \left[\left(S_{T_1} - S_{T_0} \right)_+ \middle| \mathcal{F}_t \right],$$

where $\mathbb{Q}$ is a probability measure under which the discounted price of the underlying is a martingale. In particular, at time T_0, using Example 5.2.2,

$$V_{T_0} = S_{T_0} \Phi(d_1) - S_{T_0} e^{-r(T_1-T_0)} \Phi(d_2)$$

where

$$d_1 = \frac{\left(r + \frac{\sigma^2}{2} \right) (T_1 - T_0)}{\sigma \sqrt{T_1 - T_0}} \quad \text{and} \quad d_2 = \frac{\left(r - \frac{\sigma^2}{2} \right) (T_1 - T_0)}{\sigma \sqrt{T_1 - T_0}}.$$

In other words

$$
\begin{aligned}
V_{T_0} &= S_{T_0} \left\{ \Phi\left(\left(r + \frac{\sigma^2}{2} \right) \frac{\sqrt{T_1 - T_0}}{\sigma} \right) - e^{-r(T_1-T_0)} \Phi\left(\left(r - \frac{\sigma^2}{2} \right) \frac{\sqrt{T_1 - T_0}}{\sigma} \right) \right\} \\
&= c S_{T_0}
\end{aligned}
$$

where $c = c(r, \sigma, T_0, T_1)$ is independent of the asset price.

To find the price at time $t < T_0$, observe that the portfolio consisting of c units of the underlying over the time interval $0 \le t \le T_0$ exactly replicates the option at time T_0. Thus for $t < T_0$, the price is given by $c S_t$. In particular, the time zero price of the option is

$$V_0 = S_0 \left\{ \Phi\left(\left(r + \frac{\sigma^2}{2} \right) \frac{\sqrt{T_1 - T_0}}{\sigma} \right) - e^{-r(T_1-T_0)} \Phi\left(\left(r - \frac{\sigma^2}{2} \right) \frac{\sqrt{T_1 - T_0}}{\sigma} \right) \right\}.$$

$\square$

General strategy

Notice that, in order to price the forward start option, we worked our way back from time T_1. This reflects a general strategy. For a multistage option with maturity T_1 and conditions stipulated at an intermediate time T_0, we invoke the following procedure.

Valuing multistage options:
1 Find the payoff at time T_1.
2 Use Black–Scholes to value the option for $t \in [T_0, T_1]$.
3 Apply the contract conditions at time T_0.
4 Use Black–Scholes to value the option for $t \in [0, T_0]$.

We put this into action for two more examples.

Example 6.2.2 (Ratio derivative)   *A ratio derivative can be described as follows. Two times $0 < T_0 < T_1$ are fixed. The derivative matures at time T_1 when its payoff is S_{T_1}/S_{T_0}. Find the value of the option at times $t < T_1$.*

Solution: First suppose that $t \in [T_0, T_1]$. At such times S_{T_0} is known and so

$$V_t = \frac{1}{S_{T_0}} \mathbb{E}^{\mathbb{Q}} \left[e^{-r(T_1-t)} S_{T_1} \,\middle|\, \mathcal{F}_t \right]$$

where, under $\mathbb{Q}$, the discounted asset price is a martingale. Hence $V_t = S_t/S_{T_0}$. In particular, $V_{T_0} = 1$. Evidently the value of the option for $t < T_0$ is therefore $e^{-r(T_0-t)}$. □

Both forward start options and ratio derivatives, in which the strike price is set to be a function of the stock price at some intermediate time T_0, are examples of *cliquets*.

Compound options

A rather more complex class of examples is provided by the *compound options*. These are 'options on options', that is options in which the rôle of the underlying is itself played by an option. There are four basic types of compound option: call-on-call, call-on-put, put-on-call and put-on-put.

Example 6.2.3 (Call-on-call option)   *To describe the call-on-call option we must specify two exercise prices, K_0 and K_1, and two maturity times $T_0 < T_1$. The 'underlying' option is a European call with strike price K_1 and maturity T_1. The call-on-call contract gives the holder the right to buy the underlying option for price K_0 at time T_0. Find the value of such an option for $t < T_0$.*

Solution: We know how to price the underlying call. Its value at time T_0 is given by the Black–Scholes formula as

$$\begin{aligned} C\left(S_{T_0}, T_0; K_1, T_1\right) &= S_{T_0} \Phi\left(d_1\left(S_{T_0}, T_1 - T_0, K_1\right)\right) \\ &\quad - K_0 e^{-r(T_1-T_0)} \Phi\left(d_2\left(S_{T_0}, T_1 - T_0, K_1\right)\right) \end{aligned}$$

where

$$d_1\left(S_{T_0}, T_1 - T_0, K_1\right) = \frac{\log\left(\frac{S_{T_0}}{K_1}\right) + \left(r + \frac{\sigma^2}{2}\right)(T_1 - T_0)}{\sigma\sqrt{T_1 - T_0}}$$

and $d_2\left(S_{T_0}, T_1 - T_0, K_1\right) = d_1\left(S_{T_0}, T_1 - T_0, K_1\right) - \sigma\sqrt{T_1 - T_0}$. The value of the compound option at time T_0 is then

$$V\left(T_0, S_{T_0}\right) = \left(C\left(S_{T_0}, T_0; K_1, T_1\right) - K_0\right)_+ .$$

Now we apply Black–Scholes again. The value of the option at times $t < T_0$ is

$$V(t, S_t) = e^{-r(T_0-t)} \mathbb{E}^{\mathbb{Q}}\left[\left(C(S_{T_0}, T_0, K_1, T_1) - K_0\right)_+ \middle| \mathcal{F}_t^S\right] \tag{6.3}$$

where the discounted asset price is a martingale under $\mathbb{Q}$. Using that

$$S_{T_0} = S_t \exp\left(\sigma Z\sqrt{T_0 - t} + \left(r - \frac{1}{2}\sigma^2\right)(T_0 - t)\right),$$

where, under $\mathbb{Q}$, $Z \sim N(0, 1)$, equation (6.3) now gives an analytic expression for the value in terms of the cumulative distribution function of a bivariate normal random variable. We write

$$f(y) = S_0 \exp\left(\sigma y \sqrt{T_0 - t} + \left(r - \frac{1}{2}\sigma^2\right)(T_0 - t)\right)$$

and define x_0 implicitly by

$$x_0 = \inf\{y \in \mathbb{R} : C(f(y), T_0; K_1, T_1) \geq K_0\}.$$

Now

$$\log\left(\frac{f(y)}{K_1}\right) = \log\left(\frac{S_0}{K_1}\right) + \sigma y \sqrt{T_0 - t} + \left(r - \frac{1}{2}\sigma^2\right)(T_0 - t)$$

and so writing

$$\hat{d}_1(y) = \frac{\log(S_0/K_1) + \sigma y \sqrt{T_0 - t} + rT_1 - \sigma^2 T_0 + \frac{1}{2}\sigma^2 T_1}{\sigma\sqrt{T_1 - T_0}}$$

and

$$\hat{d}_2(y) = \frac{\log(S_0/K_1) + \sigma y \sqrt{T_0 - t} + rT_1 - \frac{1}{2}\sigma^2 T_1}{\sigma\sqrt{T_1 - T_0}}$$

we obtain

$$V(t, S_t) = e^{-r(T_0 - t)} \int_{x_0}^{\infty} \left(f(y)\Phi(\hat{d}_1(y)) - K_0 e^{-r(T_1 - T_0)}\Phi(\hat{d}_2(y)) - K_0\right)$$
$$\times \frac{1}{\sqrt{2\pi}} e^{-y^2/2} dy.$$

$\square$

6.3 Lookbacks and barriers

We now turn to our first example of path-dependent options, that is options for which the history of the asset price over the duration of the contract determines the payout at expiry.

As usual we use $\{S_t\}_{0 \leq t \leq T}$ to denote the price of the underlying asset over the duration of the contract. In this section we shall consider options whose payoff at maturity depends on S_T and one or both of the maximum and minimum values taken by the asset price over $[0, T]$.

Notation: We write

$$S_*(t) = \min\{S_u : 0 \leq u \leq t\},$$

$$S^*(t) = \max\{S_u : 0 \leq u \leq t\}.$$

Definition 6.3.1 (Lookback call)   A lookback call *gives the holder the right to buy a unit of stock at time T for a price equal to the minimum achieved by the stock up to time T. That is the payoff is*

$$C_T = S_T - S_*(T).$$

Definition 6.3.2 (Barrier options)   A barrier option *is one that is activated or deactivated if the asset price crosses a preset barrier. There are two basic types:*

1 **knock-ins**

   (a) *the barrier is* up-and-in *if the option is only active if the barrier is hit from below,*
   (b) *the barrier is* down-and-in *if the option is only active if the barrier is hit from above;*

2 **knock-outs**

   (a) *the barrier is* up-and-out *if the option is worthless if the barrier is hit from below,*
   (b) *the barrier is* down-and-out *if the option is worthless if the barrier is hit from above.*

Example 6.3.3   A down-and-in call option *pays out* $(S_T - K)_+$ *only if the stock price fell below some preagreed level c some time before T, otherwise it is worthless. That is, the payoff is*

$$C_T = \mathbf{1}_{\{S_*(T) \leq c\}}(S_T - K)_+.$$

As always we can express the value of such an option as a discounted expected value under the martingale measure $\mathbb{Q}$. Thus the value at time zero can be written as

$$V(0, S_0) = e^{-rT}\mathbb{E}^{\mathbb{Q}}[C_T] \tag{6.4}$$

where r is the riskless borrowing rate and the discounted stock price is a $\mathbb{Q}$-martingale. However, in order to actually evaluate the expectation in (6.4) for barrier options we need to know the joint distribution of $(S_T, S_*(T))$ and $(S_T, S^*(T))$ under the martingale measure $\mathbb{Q}$. Fortunately we did most of the work in Chapter 3.

Joint distribution of the stock price and its minimum

In Lemma 3.3.4 we found the joint distribution of Brownian motion and its maximum. Specifically, if $\{W_t\}_{t \geq 0}$ is a standard $\mathbb{P}$-Brownian motion, writing $M_t = \max_{0 \leq s \leq t} W_s$, for $a > 0$ and $x \leq a$

$$\mathbb{P}[M_t \geq a, W_t \leq x] = 1 - \Phi\left(\frac{2a - x}{\sqrt{t}}\right).$$

By symmetry, writing $m_t = \min_{0 \leq s \leq t} W_s$, for $a < 0$ and $x \geq a$,

$$\mathbb{P}[m_t \leq a, W_t \geq x] = 1 - \Phi\left(\frac{-2a + x}{\sqrt{t}}\right),$$

or, differentiating, if $a < 0$ and $x \geq a$

$$\mathbb{P}[m_T \leq a, W_T \in dx] = p_T(0, -2a + x)dx = p_T(2a, x)dx$$

where

$$p_t(x, y) = \frac{1}{\sqrt{2\pi t}} \exp\left(-|x - y|^2/2t\right).$$

Combining these results with (two applications of) the Girsanov Theorem will allow us to calculate the joint distribution of $(S_T, S^*(T))$ and of $(S_T, S_*(T))$ under the martingale measure $\mathbb{Q}$.

As usual, under the market measure $\mathbb{P}$,

$$S_t = S_0 \exp\left(\nu t + \sigma W_t\right)$$

where $\{W_t\}_{t \geq 0}$ is a $\mathbb{P}$-Brownian motion. Let us suppose, temporarily, that $\nu = 0$ so that $S_t = S_0 \exp(\sigma W_t)$ and moreover $S_*(t) = S_0 \exp(\sigma m_t)$ and $S^*(t) = S_0 \exp(\sigma M_t)$. In this special case then the joint distribution of the stock price and its minimum (resp. maximum) can be deduced from that of (W_t, m_t) (resp. (W_t, M_t)). Of course, in general, ν will not be zero either under the market measure $\mathbb{P}$ or under the martingale measure $\mathbb{Q}$. Our strategy will be to use the Girsanov Theorem not only to switch to the martingale measure but also to switch, temporarily, to an equivalent measure under which $S_t = S_0 \exp(\sigma W_t)$.

Lemma 6.3.4   Let $\{Y_t\}_{t \geq 0}$ be given by $Y_t = bt + X_t$ where b is a constant and $\{X_t\}_{t \geq 0}$ is a $\mathbb{Q}$-Brownian motion. Writing $Y_*(t) = \min\{Y_u : 0 \leq u \leq t\}$,

$$\mathbb{Q}[Y_*(T) \leq a, Y_T \in dx] = \begin{cases} p_T(bT, x)dx & \text{if } x < a, \\ e^{2ab} p_T(2a + bT, x)dx & \text{if } x \geq a, \end{cases}$$

where, as above, $p_t(x, y)$ is the Brownian transition density function.

Proof: By the Girsanov Theorem, there is a measure $\mathbb{P}$, equivalent to $\mathbb{Q}$, under which $\{Y_t\}_{t \geq 0}$ is a $\mathbb{P}$-Brownian motion and

$$\left.\frac{d\mathbb{P}}{d\mathbb{Q}}\right|_{\mathcal{F}_r} = \exp\left(-bX_T - \frac{1}{2}b^2 T\right).$$

Notice that this depends on $\{X_t\}_{0 \leq t \leq T}$ only through X_T. The $\mathbb{Q}$-probability of the event $\{Y_*(T) \leq a, Y_T \in dx\}$ will be the $\mathbb{P}$-probability of that event multiplied by $\left.\frac{d\mathbb{Q}}{d\mathbb{P}}\right|_{\mathcal{F}_r}$ evaluated at $Y_T = x$. Now

$$\frac{d\mathbb{Q}}{d\mathbb{P}} = \exp\left(bX_T + \frac{1}{2}b^2 T\right) = \exp\left(bY_T - \frac{1}{2}b^2 T\right)$$

and so for $a < 0$ and $x \geq a$

$$\begin{aligned} \mathbb{Q}[Y_*(T) \leq a, Y_T \in dx] &= \mathbb{P}[Y_*(T) \leq a, Y_T \in dx]\exp\left(bx - \frac{1}{2}b^2 T\right) \\ &= p_T(2a, x)\exp\left(bx - \frac{1}{2}b^2 T\right)dx \\ &= e^{2ab} p_T(2a + bT, x)dx. \end{aligned} \tag{6.5}$$

Evidently for $x \leq a$, $\{Y_*(T) \leq a, Y_T \in dx\} = \{Y_T \in dx\}$ and so for $x \leq a$

$$
\begin{aligned}
\mathbb{Q}[Y_*(T) \leq a, Y_T \in dx] &= \mathbb{Q}[Y_T \in dx] \\
&= \mathbb{Q}[bT + X_T \in dx] \\
&= p_T(bT, x)dx
\end{aligned}
$$

and the proof is complete.   □

Differentiating (6.5) with respect to a, we see that, in terms of joint densities, for $a < 0$

$$
\mathbb{Q}[Y_*(T) \in da, Y_T \in dx] = \frac{2e^{2ab}}{T}|x - 2a|p_T(2a + bT, x)dx\,da \quad \text{for } x \geq a.
$$

The joint density evidently vanishes if $x < a$ or $a > 0$. In Exercise 13 you are asked to find the joint distribution of Y_T and $Y^*(T)$ under $\mathbb{Q}$.

An expression for the price

From Chapter 5, under the martingale measure $\mathbb{Q}$, $S_t = S_0 \exp(\sigma Y_t)$ where

$$
Y_t = \frac{(r - \frac{1}{2}\sigma^2)}{\sigma}t + X_t
$$

and $\{X_t\}_{t \geq 0}$ is a $\mathbb{Q}$-Brownian motion. So by applying these results with $b = (r - \frac{1}{2}\sigma^2)/\sigma$ we can now evaluate the price of any option maturing at time T whose payoff depends just on the stock price at time T and its minimum (or maximum) value over the lifetime of the contract. If the payoff is $C_T = g(S_*(T), S_T)$ and r is the riskless borrowing rate then the value of the option at time zero is

$$
\begin{aligned}
V(0, S_0) &= e^{-rT}\mathbb{E}^{\mathbb{Q}}[g(S_*(T), S_T)] \\
&= e^{-rT}\int_{a=-\infty}^{0}\int_{x=a}^{\infty} g\left(S_0 e^{\sigma x}, S_0 e^{\sigma a}\right)\mathbb{Q}[Y_*(T) \in da, Y_T \in dx].
\end{aligned}
$$

Example 6.3.5 (Down-and-in call option)   *Find the time zero price of a down-and-in call option whose payoff at time T is*

$$
C_T = \mathbf{1}_{\{S_*(T) \leq c\}}(S_T - K)_+
$$

where c is a (positive) preagreed constant less than K.

Solution: Using $S_t = S_0 \exp(\sigma Y_t)$ we rewrite the payoff as

$$
C_T = \mathbf{1}_{\{Y_*(T) \leq \frac{1}{\sigma}\log(c/S_0)\}}\left(S_0 e^{\sigma Y_T} - K\right)_+.
$$

Writing $b = (r - \frac{1}{2}\sigma^2)/\sigma$, $a = \frac{1}{\sigma}\log(c/S_0)$ and $x_0 = \frac{1}{\sigma}\log(K/S_0)$ we obtain

$$
V(0, S_0) = e^{-rT}\int_{x_0}^{\infty}\left(S_0 e^{\sigma x} - K\right)\mathbb{Q}\left(Y_*(T) \leq a, Y_T \in dx\right).
$$

Using the expression for the joint distribution of $(Y_*(T), Y_T)$ obtained above yields

$$V(0, S_0) = e^{-rT} \int_{x_0}^{\infty} \left(S_0 e^{\sigma x} - K \right) e^{2ab} p_T(2a + bT, x) dx.$$

We have used the fact that, since $c < K$, $x_0 \geq a$. First observe that

$$e^{-rT} \int_{x_0}^{\infty} K e^{2ab} p_T(2a + bT, x) dx = K e^{-rT} e^{2ab} \int_{(x_0-2a-bT)/\sqrt{T}}^{\infty} \frac{1}{\sqrt{2\pi}} e^{-y^2/2} dy$$

$$= K e^{-rT} e^{2ab} \int_{-\infty}^{(2a+bT-x_0)/\sqrt{T}} \frac{1}{\sqrt{2\pi}} e^{-y^2/2} dy$$

$$= K e^{-rT} \left(\frac{c}{S_0} \right)^{\frac{2r}{\sigma^2}-1} \Phi\left(\frac{2a + bT - x_0}{\sqrt{T}} \right)$$

$$= K e^{-rT} \left(\frac{c}{S_0} \right)^{\frac{2r}{\sigma^2}-1} \Phi\left(\frac{\log(F/K) - \frac{1}{2}\sigma^2 T}{\sigma\sqrt{T}} \right)$$

where $F = e^{rT} c^2 / S_0$.

Similarly,

$$e^{-rT} \int_{x_0}^{\infty} S_0 e^{\sigma x} e^{2ab} p_T(2a + bT, x) dx$$

$$= S_0 e^{-rT} e^{2ab} \int_{x_0}^{\infty} \frac{1}{\sqrt{2\pi T}} \exp\left(-\frac{(x - (2a + bT))^2 - 2\sigma x T}{2T} \right) dx$$

$$= S_0 e^{-rT} e^{2ab} \int_{(x_0-(2a+bT)-\sigma T)/\sqrt{T}}^{\infty} \frac{1}{\sqrt{2\pi}} e^{-y^2/2} dy$$

$$\times \exp\left(\frac{1}{2}\sigma^2 T + 2a\sigma + b\sigma T \right)$$

$$= e^{-rT} \left(\frac{c}{S_0} \right)^{\frac{2r}{\sigma^2}-1} F\Phi\left(\frac{\log(F/K) + \frac{1}{2}\sigma^2 T}{\sigma\sqrt{T}} \right).$$

Comparing this with Example 5.2.2

$$V(0, S_0) = \left(\frac{c}{S_0} \right)^{\frac{2r}{\sigma^2}-1} C\left(\frac{c^2}{S_0}, 0; K, T \right),$$

where $C(x, t; K, T)$ is the price at time t of a European call option with strike K and maturity T if the stock price at time t is x.   □

The price of a barrier option can also be expressed as the solution of a partial differential equation.

Example 6.3.6 (Down-and-out call)   *A down-and-out call has the same payoff as a European call option, $(S_T - K)_+$, unless during the lifetime of the contract the price of the underlying asset has fallen below some preagreed barrier, c, in which case the option is 'knocked out' worthless.*

Writing $V(t, x)$ for the value of such an option at time t if $S_t = x$ and assuming that $K > c$, $V(t, x)$ solves the Black–Scholes equation for $(t, x) \in [0, T] \times [c, \infty)$ subject to the boundary conditions

$$V(T, S_T) = (S_T - K)_+,$$
$$V(t, c) = 0, \quad t \in [0, T],$$
$$\frac{V(t, x)}{x} \to 1, \quad as\ x \to \infty.$$

The last boundary condition follows since as $S_t \to \infty$, the probability of the asset price hitting level c before time T tends to zero.

Exercise 16 provides a method for solving the Black–Scholes partial differential equation with these boundary conditions.

Of course more and more complicated barrier options can be dreamt up. For example, a *double knock-out option* is worthless if the stock price leaves some interval $[c_1, c_2]$ during the lifetime of the contract. The probabilistic pricing formula for such a contract then requires the joint distribution of the triple $(S_T, S_*(T), S^*(T))$. As in the case of a single barrier, the trick is to use Girsanov's Theorem to deduce the joint distribution from that of (W_T, m_T, M_T) where $\{W_t\}_{t \geq 0}$ is a $\mathbb{P}$-Brownian motion and $\{m_t\}_{t \geq 0}$, $\{M_t\}_{t \geq 0}$ are its running minimum and maximum respectively. This in turn is given by

$$\mathbb{P}[W_T \in dy, a < m_T, M_T < b] = \sum_{n \in \mathbb{Z}} \left\{ p_T(2n(a-b), y) - p(2n(b-a), y-2a) \right\} dy;$$

see Freedman (1971) for a proof. An explicit pricing formula will then be in the form of an infinite sum. In Exercise 20 you obtain the pricing formula by directly solving the Black–Scholes differential equation.

Probability or pde? As we have seen in Exercise 7 of Chapter 5 and we see again in the exercises at the end of this chapter, the Black–Scholes partial differential equation can be solved by first transforming it to the heat equation (with appropriate boundary conditions). This is entirely parallel to our probabilistic technique of transforming the expectation price to an expectation of a function of Brownian motion.

6.4 Asian options

The payoff of an Asian option is a function of the average of the asset price over the lifetime of the contract. For example, the payoff of an *Asian call* with strike price K and maturity time T is

$$C_T = \left(\frac{1}{T} \int_0^T S_t dt - K \right)_+.$$

Evidently $C_T \in \mathcal{F}_T$ and so our Black–Scholes analysis of Chapter 5 gives the value of such an option at time zero as

$$V_0 = \mathbb{E}^{\mathbb{Q}} \left[e^{-rT} \left(\frac{1}{T} \int_0^T S_t dt - K \right)_+ \right]. \tag{6.6}$$

However, evaluation of this integral is a highly non-trivial matter and we do not obtain the nice explicit formulae of the previous sections.

There are many variants on this theme. For example, we might want to value a claim with payoff

$$C_T = f\left(S_T, \frac{1}{T}\int_0^T S_t dt\right).$$

In §7.2 we shall develop the technology to express the price of such claims (and indeed slightly more complex claims) as solutions to a multidimensional version of the Black–Scholes equation. Moreover (see Exercise 12 of Chapter 7) one can also find an explicit expression for the hedging portfolio in terms of the solution to this equation. However, multidimensional versions of the Black–Scholes equation are much harder to solve than their one-dimensional counterpart and generally one must resort to numerical techniques.

The main difficulty with evaluating (6.6) directly is that, although there are explicit formulae for all the moments of the average process $\frac{1}{T}\int_0^T S_t dt$, in contrast to the lognormal distribution of S_T, we do not have an expression for the distribution function. A number of approaches have been suggested to overcome this, including simply approximating the distribution of the average process by a lognormal distribution with suitably chosen parameters.

A very natural approach is to replace the continuous average by a discrete analogue obtained by sampling the price of the process at agreed times $t_1, \ldots, t_n$ and averaging the result. This also makes sense from a practical point of view as calculating the continuous average for a real asset can be a difficult process. Many contracts actually specify that the average be calculated from such a discrete sample – for example from daily closing prices. Mathematically, the continuous average $\frac{1}{T}\int_0^T S_t dt$ is replaced by $\frac{1}{n}\sum_{i=1}^n S_{t_i}$. Options based on a discrete sample can be treated in the same way as multistage options, although evaluation of the price rapidly becomes impractical (see Exercise 21).

A further approximation is to replace the arithmetic average by a *geometric* average. That is, in place of $\frac{1}{n}\sum_{i=1}^n S_{t_i}$ we consider $\left(\prod_{i=1}^n S_{t_i}\right)^{1/n}$. This quantity has a lognormal distribution (Exercise 22) and so the corresponding approximate pricing formula for the Asian option can be evaluated exactly. (You are asked to find the pricing formula for an Asian call option based on a continuous version of the geometric average in Exercise 23.) Of course the arithmetic mean of a collection of positive numbers always dominates their geometric mean and so it is no surprise that this approximation consistently under-prices the Asian call option.

6.5   American options

A full treatment of American options is beyond our scope here. Explicit formulae for the prices of American options only exist in a few special cases and so one must employ numerical techniques. One approach is to use our discrete (binomial tree) models of Chapter 2. An alternative is to reformulate the price as a solution to a partial differential equation. We do not give a rigorous derivation of this equation, but instead we use the results of Chapter 2 to give a heuristic explanation of its form.

The discrete case

As we saw in Chapter 2, the price of an American call option on non-dividend-paying stock is the same as that of a European call and so we concentrate on the *American put*. This option gives the holder the right to buy one unit of stock for price K at any time before the maturity time T.

As we illustrated in §2.2, in our discrete time model, if $V(n, S_n)$ is the value of the option at time $n\delta t$ given that the asset price at time $n\delta t$ is S_n then

$$V(n, S_n) = \max \left\{ (K - S_n)_+, \mathbb{E}^{\mathbb{Q}} \left[e^{-r\delta t} V(n+1, S_{n+1}) \big| \mathcal{F}_n \right] \right\},$$

where $\mathbb{Q}$ is the martingale measure. In particular, $V(n, S_n) \geq (K - S_n)_+$ everywhere. We saw that for each fixed n the possible values of S_n are separated into two ranges by a boundary value that we shall denote by $S_f(n)$: if $S_n > S_f(n)$ then it is optimal to hold the option whereas if $S_n \leq S_f(n)$ it is optimal to exercise. We call $\{S_f(n)\}_{0 \leq n \leq N}$ the *exercise boundary*.

In Example 2.4.7 we found a characterisation of the exercise boundary. We showed that the discounted option price can be written as $\tilde{V}_n = \tilde{M}_n - \tilde{A}_n$ where $\{\tilde{M}_n\}_{0 \leq n \leq N}$ is a $\mathbb{Q}$-martingale and $\{\tilde{A}_n\}_{0 \leq n \leq N}$ is a non-decreasing predictable process. The option is exercised at the first time $n\delta t$ when $\tilde{A}_{n+1} \neq 0$. In summary, within the exercise region $\tilde{A}_{n+1} \neq 0$ and $V_n = (K - S_n)_+$, whereas away from the exercise region, that is when $S_n > S_f(n)$, $V(n, S_n) = M_n$.

The strategy of exercising the option at the first time when $\tilde{A}_{n+1} \neq 0$ is *optimal* in the sense that if we write $\mathcal{T}_N$ for the set of all possible stopping times taking values in $\{0, 1, \ldots, N\}$ then

$$V(0, S_0) = \sup_{\tau \in \mathcal{T}_N} \mathbb{E}^{\mathbb{Q}} \left[e^{-r\tau} (K - S_\tau)_+ \big| \mathcal{F}_0 \right].$$

Since the exercise time of any permissible strategy must be a stopping time, this says that as holder of the option one can't do better by choosing any other exercise strategy. That this optimality characterises the fair price follows from a now familiar arbitrage argument that you are asked to provide in Exercise 24.

Continuous time

Now suppose that we formally pass to the continuous limit as in §2.6. We expect that in the limit too $V(t, S_t) \geq (K - S_t)_+$ everywhere and that for each t we can define $S_f(t)$ so that if $S_t > S_f(t)$ it is optimal to hold on to the option, whereas if $S_t \leq S_f(t)$ it is optimal to exercise. In the exercise region $V(t, S_t) = (K - S_t)_+$ whereas away from the exercise region $V(t, S_t) = M_t$ where the discounted process $\{\tilde{M}_t\}_{0 \leq t \leq T}$ is a $\mathbb{Q}$-martingale and $\mathbb{Q}$ is the measure, equivalent to $\mathbb{P}$, under which the discounted stock price is a martingale. Since $\{\tilde{M}_t\}_{0 \leq t \leq T}$ can be thought of as the discounted value of a European option, this tells us that away from the exercise region, $V(t, x)$ must satisfy the Black–Scholes differential equation.

We guess then that for $\{(t, x) : x > S_f(t)\}$ the price $V(t, x)$ must satisfy the Black–Scholes equation whereas outside this region $V(t, x) = (K - x)_+$. This can be extended to a characterisation of $V(t, x)$ if we specify appropriate boundary conditions on S_f. This is complicated by the fact that $S_f(t)$ is a *free boundary* – we don't know its location a priori.

An arbitrage argument (Exercise 25) says that the price of an American put option should be continuous. We have checked already that $V(t, S_f(t)) = (K - S_f(t))_+$. Since it is clearly not optimal to exercise at a time $t < T$ if the value of the option is zero, in fact we have $V(t, S_f(t)) = K - S_f(t)$. Let us suppose now that $V(t, x)$ is continuously differentiable with respect to x as we cross the exercise boundary (we shall omit the proof of this). Then, since

$$V(t, x) = (K - x) \quad \text{for } x \leq S_f \text{ and}$$
$$V(t, x) \geq (K - x) \quad \text{for } x > S_f,$$

we must have that at the exercise boundary $\frac{\partial V}{\partial x} \geq -1$. Suppose that $\frac{\partial V}{\partial x} > -1$ at some point of the exercise boundary. Then by reducing the value of the stock price at which we choose to exercise from S_f to S_f^* we can actually *increase* the value of the option at $(t, S_f(t))$. This contradicts the optimality of our exercise strategy. It must be that $\frac{\partial V}{\partial x} = -1$ at the exercise boundary.

We can now fully characterise $V(t, x)$ as a solution to a free boundary value problem:

Proposition 6.5.1 (The value of an American put)   *We write $V(t, x)$ for the value of an American put option with strike price K and maturity time T and r for the riskless borrowing rate. $V(t, x)$ can be characterised as follows. For each time $t \in [0, T]$ there is a number $S_f(t) \in (0, \infty)$ such that for $0 \leq x \leq S_f(t)$ and $0 \leq t \leq T$,*

$$V(t, x) = K - x \quad and \quad \frac{\partial V}{\partial t} + \frac{1}{2}\sigma^2 x^2 \frac{\partial^2 V}{\partial x^2} + rx\frac{\partial V}{\partial x} - rV < 0.$$

For $t \in [0, T]$ and $S_f(t) < x < \infty$

$$V(t, x) > (K - x)_+ \quad and \quad \frac{\partial V}{\partial t} + \frac{1}{2}\sigma^2 x^2 \frac{\partial^2 V}{\partial x^2} + rx\frac{\partial V}{\partial x} - rV = 0.$$

The boundary conditions at $x = S_f(t)$ are that the option price process is continuously differentiable with respect to x, is continuous in time and

$$V(t, S_f(t)) = (K - S_f(t))_+, \quad \frac{\partial V}{\partial x}(t, S_f(t)) = -1.$$

In addition, V satisfies the terminal condition

$$V(T, S_T) = (K - S_T)_+.$$

The free boundary problem of Proposition 6.5.1 is easier to analyse as a *linear complementarity problem*. If we use the notation

$$\mathcal{L}_{BS}f = \frac{\partial f}{\partial t} + \frac{1}{2}\sigma^2 x^2 \frac{\partial^2 f}{\partial x^2} + rx\frac{\partial f}{\partial x} - rf,$$

then the free boundary value problem can be restated as

$$\mathcal{L}_{BS}V(t, x)\,(V(t, x) - (K - x)_+) = 0,$$

subject to $\mathcal{L}_{BS}V(t, x) \le 0$, $V(t, x) - (K - x)_+ \ge 0$, $V(T, x) = (K - x)_+$, $V(t, x) \to \infty$ as $x \to \infty$ and $V(t, x)$, $\frac{\partial V}{\partial x}(t, x)$ are continuous.

Notice that this reformulation has removed explicit dependence on the free boundary. Variational techniques can be applied to solve the problem and then the boundary is recovered from that solution. This is beyond our scope here. See Wilmott, Howison & Dewynne (1995) for more detail.

An explicit solution

We finish this chapter with one of the rare examples of an American option for which the price can be obtained explicitly.

Example 6.5.2 (Perpetual American put)   *Find the value of a perpetual American put option on non-dividend-paying stock, that is a contract that the holder can choose to exercise at any time t in which case the payoff is $(K - S_t)_+$.*

Solution(s): We sketch *two* possible solutions to this problem, first via the free boundary problem of Proposition 6.5.1 and second via the expectation price.

Since the time to expiry of the contract is always infinite, $V(t, x)$ is a function of x alone and the exercise boundary must be of the form $S_f(t) = \alpha$ for *all* $t > 0$ and some constant α. The option will be exercised as soon as $S_t \le \alpha$. The Black–Scholes equation reduces to an *ordinary* differential equation:

$$\frac{1}{2}\sigma^2 x^2 \frac{d^2 V}{dx^2} + rx \frac{dV}{dx} - rV = 0, \quad \text{for all } x \in (\alpha, \infty). \tag{6.7}$$

The general solution to equation (6.7) is of the form $v(x) = c_1 x^{d_1} + c_2 x^{d_2}$ for some constants c_1, c_2, d_1 and d_2. Fitting the boundary conditions

$$V(\alpha) = K - \alpha, \quad \lim_{x \downarrow \alpha} \frac{dV}{dx} = -1 \quad \text{and} \quad \lim_{x \to \infty} V(x) = 0$$

gives

$$V(x) = \begin{cases} (K - \alpha)\left(\frac{\alpha}{x}\right)^{2r\sigma^{-2}}, & x \in (\alpha, \infty), \\ (K - x), & x \in [0, \alpha], \end{cases}$$

where

$$\alpha = \frac{2r\sigma^{-2}K}{2r\sigma^{-2} + 1}.$$

An alternative approach to this problem would be to apply the results of §3.3. As we argued above, the option will be exercised when the stock price first hits level α for some $\alpha > 0$. This means that the value will be of the form

$$V(0, S_0) = \mathbb{E}^{\mathbb{Q}}\left[e^{-r\tau_\alpha}(K - \alpha)_+\right],$$

where $\tau_\alpha = \inf\{t > 0 : S_t \le \alpha\}$. We rewrite this stopping time in terms of the time that it takes a $\mathbb{Q}$-Brownian motion to hit a sloping line. Since

$$S_t = S_0 \exp\left(\left(r - \frac{1}{2}\sigma^2\right)t + \sigma X_t\right)$$

where $\{X_t\}_{t\ge 0}$ is a standard Brownian motion under the martingale measure $\mathbb{Q}$, the event $\{S_t \le \alpha\}$ is the same as the event

$$\left\{-\sigma X_t - \left(r - \frac{1}{2}\sigma^2\right)t \ge \log\left(\frac{S_0}{\alpha}\right)\right\}.$$

The process $\{-X_t\}_{t\ge 0}$ is also a standard $\mathbb{Q}$-Brownian motion and so, in the notation of §3.3, the time τ_α is given by $T_{a,b}$ with

$$a = \frac{1}{\sigma}\log\left(\frac{S_0}{\alpha}\right), \quad b = \frac{r - \frac{1}{2}\sigma^2}{\sigma}.$$

We can then read off $\mathbb{E}^{\mathbb{Q}}\left[e^{-r\tau_\alpha}\right]$ from Proposition 3.3.5 and maximise over α to yield the result. $\qquad\square$

Exercises

1 Let K_1 and K_2 be fixed real numbers with $0 < K_1 < K_2$. A *collar option* has payoff

$$C_T = \min\{\max\{S_T, K_1\}, K_2\}.$$

Find the Black–Scholes price for such an option.

2 What is the maximum potential loss associated with taking the long position in a forward contract? And with taking the short position?

Consider the derivative whose payoff at expiry to the holder of the long position is

$$C_T = \min\{S_T, F\} - K,$$

where F is the standard forward price for the underlying stock and K is a constant. Such a contract is constructed so as to have zero value at the time at which it is struck. Find an expression for the value of K that should be written into such a contract. What is the maximum potential loss for the holder of the long or short position now?

3 The *digital put option* with strike K at time T has payoff

$$C_T = \begin{cases} 0, & S_T \ge K, \\ 1, & S_T < K. \end{cases}$$

Find the Black–Scholes price for a digital put. What is the put–call parity for digital options?

4   *Digital call option* In Example 6.1.1 we calculated the price of a digital call. Here is
    an alternative approach:

    (a) Use the Feynman–Kac stochastic representation to find the partial differential
        equation satisfied by the value of a digital call with strike K and maturity T.
    (b) Show that the delta of a standard European call option solves the partial
        differential equation that you have found in (a).
    (c) Hence or otherwise solve the equation in (a) to find the value of the digital.

5   An *asset-or-nothing* call option with strike K and maturity T has payoff

    $$C_T = \begin{cases} S_T, & S_T \geq K, \\ 0, & S_T < K. \end{cases}$$

    Find the Black–Scholes price and hedge for such an option. What happens to the
    stock holding in the replicating portfolio if the asset price is near K at times close to
    T? Comment.

6   Construct a portfolio consisting entirely of cash-or-nothing and asset-or-nothing
    options whose value at time T is exactly that of a European call option with strike K
    at maturity T.

7   In §6.1 we have seen that for certain options with discontinuous payoffs at maturity,
    the stock holding in the replicating portfolio can oscillate wildly close to maturity.
    Do you see this phenomenon if the payoff is continuous?

8   *Pay-later option* This option, also known as a *contingent premium option*, is a
    standard European option except that the buyer pays the premium only at maturity of
    the option and then only if the option is in the money. The premium is chosen so that
    the value of the option at time zero is zero. This option is equivalent to a portfolio
    consisting of one standard European call option with strike K and maturity T and
    $-V$ digital call options with maturity T where V is the premium for the option.

    (a) What is the value of holding such a portfolio at time zero?
    (b) Find an expression for V.
    (c) If a speculator enters such a contract, what does this suggest about her market
        view?

9   *Ratchet option* A two-leg ratchet call option can be described as follows. At time
    zero an initial strike price K is set. At time $T_0 > 0$ the strike is *reset* to S_{T_0}, the value
    of the underlying at time T_0. At the maturity time $T_1 > T_0$ the holder receives the
    payoff of the call with strike S_{T_0} plus $S_{T_1} - S_{T_0}$ if this is positive. That is, the payoff
    is $(S_{T_1} - S_{T_0})_+ + (S_{T_0} - K)_+$.
    If $(S_{T_0} - K)$ is positive, then the intermediate profit $(S_{T_0} - K)_+$ is said to be 'locked
    in'. Why? Value this option for $0 < t < T_1$.

10 *Chooser option* A chooser option is specified by two strike prices, K_0 and K_1, and two maturity dates, $T_0 < T_1$. At time T_0 the holder has the right to buy, for price K_0, *either* a call *or* a put with strike K_1 and maturity T_1.

What is the value of the option at time T_0? In the special case $K_0 = 0$ use put–call parity to express this as the sum of the value of a call and a put with suitably chosen strike prices and maturity dates and hence find the value of the option at time zero.

11 *Options on futures* In our simple model where the riskless rate of borrowing is deterministic, forward and futures prices coincide. A European call option with strike price K and maturity T_0 written on an underlying futures contract with delivery date $T_1 > T_0$ delivers to the holder, at time T_0, a long position in the futures contract and an amount of money $(F(T_0, T_1) - K)_+$, where $F(T_0, T_1)$ is the value of the futures contract at time T_0. Find the value of such an option at time zero.

12 Use the method of Example 6.2.3 to find the value of a put-on-put option.

By considering the portfolio obtained by buying one call-on-put and selling one put-on-put (with the same strikes and maturities) obtain a put–call parity relation for compound options. Hence write down prices for all four classes of compound option.

13 Let $\{Y_t\}_{t \geq 0}$ be given by $Y_t = bt + X_t$ where b is a constant and $\{X_t\}_{t \geq 0}$ is a $\mathbb{Q}$-Brownian motion. Writing $Y^*(t) = \max\{Y_u : 0 \leq u \leq t\}$, find the joint distribution of $(Y_T, Y^*(T))$ under $\mathbb{Q}$.

14 What is the value of a portfolio consisting of one down-and-in call and one down-and-out call with the same strike price and maturity?

15 Find the value of a down-and-out call with barrier c and strike K at maturity T if $c > K$.

16 One approach to finding the value of the down-and-out call of Example 6.3.6 is to express it as an expectation under the martingale measure and exploit our knowledge of the joint distribution of Brownian motion and its minimum. Alternatively one can solve the partial differential equation directly and that is the purpose of this exercise.

(a) Use the method of Exercise 7 of Chapter 5 to transform the equation for the price into the heat equation. What are the boundary conditions for this heat equation?

(b) Solve the heat equation that you have obtained using, for example, the 'method of images'. (If you are unfamiliar with this technique, then try Wilmott, Howison & Dewynne (1995).)

(c) Undo the transformation to obtain the solution to the partial differential equation.

17 An *American cash-or-nothing call option* can be exercised at any time $t \in [0, T]$. If exercised at time t its payoff is

$$
\begin{array}{ll}
1 & \text{if } S_t \geq K, \\
0 & \text{if } S_t < K.
\end{array}
$$

When will such an option be exercised? Find its value.

18 Suppose that the down-and-in call option of Example 6.3.5 is modified so that if the option is never activated, that is the stock price never crosses the barrier, then the holder receives a rebate of Z. Find the price of this modified option.

19 A *perpetual option* is one with no expiry time. For example, a perpetual American cash-or-nothing call option can be exercised at any time. If exercised at time t, its payoff is 1 if $S_t \geq K$ and 0 if $S_t < K$. What is the probability that such an option is never exercised?

20 Formulate the price of a double knock-out call option as a solution to a partial differential equation with suitably chosen boundary conditions. Mimic your approach in Exercise 16 to see that this too leads to an expression for the price as an infinite sum.

21 Calculate the value of an Asian call option, with strike price K, in which the average of the stock price is calculated on the basis of just two sampling times, 0 and T, where T is the maturity time of the contract.
   Find an expression for the value of the corresponding contract when there are three sampling times, 0, $T/2$ and T.

22 Suppose that $\{S_t\}_{t \geq 0}$ is a geometric Brownian motion under $\mathbb{P}$. Let $0 \leq t_1 \leq t_2 \leq \cdots \leq t_n$ be fixed times and define

$$
G_n = \left(\prod_{i=1}^{n} S_{t_i} \right)^{1/n}.
$$

Show that G_n has a lognormal distribution under $\mathbb{P}$.

23 An asset price $\{S_t\}_{t \geq 0}$ is a geometric Brownian motion under the market measure $\mathbb{P}$. Define

$$
Y_T = \exp\left(\frac{1}{T} \int_0^T \log S_t \, dt \right).
$$

Suppose that an Asian call option has payoff $(Y_T - K)_+$ at time T. Find an explicit formula for the price of such an option at time zero.

24 Use an arbitrage argument to show that if $V(0, S_0)$ is the fair price of an American put option on non-divident-paying stock with strike price K and maturity T, then writing $\mathcal{T}_T$ for the set of all possible stopping times taking values in $[0, T]$

$$
V(0, S_0) = \sup_{\tau \in \mathcal{T}_T} \mathbb{E}^{\mathbb{Q}} \left[e^{-r\tau} (K - S_\tau)_+ \big| \mathcal{F}_0 \right].
$$

25   Consider the value of an American put on non-dividend-paying stock. Show that
     if there were a discontinuity in the option value (as a function of stock price) that
     persisted for more than an infinitesimal time then a portfolio consisting entirely of
     options would offer an arbitrage opportunity.
     Remark: This does not mean that *all* option prices are continuous. If there is an
     instantaneous change in the conditions of a contract (as in multistage options) then
     discontinuities certainly can occur.

26   Find the value of a perpetual American call option on non-dividend-paying stock.

7 Bigger models

Summary

Having applied our basic Black–Scholes model to the pricing of some exotic options, we now turn to more general market models.

In §7.1 we replace the (constant) parameters that characterised our basic Black–Scholes model by previsible processes. Under appropriate boundedness assumptions, we then repeat our analysis of Chapter 5 to obtain the fair price of an option as the discounted expected value of the claim under a martingale measure. In general this expectation must be evaluated numerically. We also make the connection with a generalised Black–Scholes equation via the Feynman–Kac Stochastic Representation Theorem.

Our models so far have assumed that the market consists of a single stock and a riskless cash bond. More complex equity products can depend on the behaviour of several separate securities and, in general, the prices of these securities will not evolve independently. In §7.2 we extend some of the fundamental results of Chapter 4 to allow us to manipulate systems of stochastic differential equations driven by *correlated* Brownian motions. For markets consisting of many assets we have much more freedom in our choice of 'reference asset' or numeraire and so we revisit this issue before illustrating the application of the 'multifactor' theory by pricing a 'quanto' product.

We still have no satisfactory justification for the geometric Brownian motion model. Indeed, there is considerable evidence that it does not capture all features of stock price evolution. A first objection is that stock prices occasionally 'jump' at unpredictable times. In §7.3 we introduce a Poisson process of jumps into our Black–Scholes model and investigate the implications for option pricing. This approach is popular in the analysis of credit risk. In §1.5 we saw that, if a model is to be free from arbitrage and complete, there must be a balance between the number of sources of randomness and the number of independent stocks. We reiterate this here. We see more evidence that the Black–Scholes model does not reflect the true behaviour of the market in §7.4. It seems a little late in the day to condemn the model that has been the subject of all our efforts so far and so we ask how much it matters if we use the wrong model. We also very briefly discuss models with stochastic volatility that have the potential to better reflect true market behaviour.

This chapter is intended to do no more than indicate some of the topics that might be addressed in a *second* course in financial calculus. Much more detail can be found in some of the suggestions for further reading in the bibliography.

7.1   General stock model

In our classical Black–Scholes framework we assume that the riskless borrowing rate is constant and that the returns of the stock follow a Brownian motion with constant drift. In this section we consider much more general models to which we can apply the Black–Scholes analysis although, in practice, even for vanilla options the prices that we obtain must now be evaluated numerically. The key assumption that we retain is that there is only one source of randomness in the market, the Brownian motion that drives the stock price (cf. §7.3).

The model   Writing $\{\mathcal{F}_t\}_{t\geq0}$ for the filtration generating the driving Brownian motion, we replace the riskless borrowing rate, r, the drift μ and the volatility σ in our basic Black–Scholes model by $\{\mathcal{F}_t\}_{t\geq0}$-predictable processes $\{r_t\}_{t\geq0}$, $\{\mu_t\}_{t\geq0}$ and $\{\sigma_t\}_{t\geq0}$. In particular, r_t, μ_t and σ_t can depend on the whole history of the market before time t. Our market model is then as follows.

General stock model: The market consists of a riskless cash bond, $\{B_t\}_{t\geq0}$, and a single risky asset with price process $\{S_t\}_{t\geq0}$ governed by

$$dB_t = r_t B_t dt, \qquad B_0 = 1,$$
$$dS_t = \mu_t S_t dt + \sigma_t S_t dW_t,$$

where $\{W_t\}_{t\geq0}$ is a $\mathbb{P}$-Brownian motion generating the filtration $\{\mathcal{F}_t\}_{t\geq0}$ and $\{r_t\}_{t\geq0}$, $\{\mu_t\}_{t\geq0}$ and $\{\sigma_t\}_{t\geq0}$ are $\{\mathcal{F}_t\}_{t\geq0}$-predictable processes.

Evidently a solution to these equations should take the form

$$B_t = \exp\left(\int_0^t r_s ds\right), \tag{7.1}$$

$$S_t = S_0 \exp\left(\int_0^t \left(\mu_s - \frac{1}{2}\sigma_s^2\right) ds + \int_0^t \sigma_s dW_s\right), \tag{7.2}$$

but we need to make some boundedness assumptions if these expressions are to make sense. So to ensure the existence of the integrals in equations (7.1) and (7.2) we assume that $\int_0^T |r_t| dt$, $\int_0^T |\mu_t| dt$ and $\int_0^T \sigma_t^2 dt$ are all finite with $\mathbb{P}$-probability one.

A word of warning is in order. In order to 'calibrate' such a model to the market we must choose the parameters $\{r_t\}_{t\geq 0}$, $\{\mu_t\}_{t\geq 0}$ and $\{\sigma_t\}_{t\geq 0}$ from an infinite-dimensional space. Unless we restrict the possible forms of these processes, this presents a major obstacle to implementation. In §7.4 we examine the effect of model misspecification on pricing and hedging strategies. Now, however, we set this worry aside and repeat the Black–Scholes analysis for our general class of market models.

A martingale measure

We must mimic the three steps to replication that we followed in the classical setting. The first of these is to find an equivalent probability measure, $\mathbb{Q}$, under which the discounted stock price, $\{\tilde{S}_t\}_{t\geq 0}$, is a martingale.

Exactly as before, we use the Girsanov Theorem to find a measure, $\mathbb{Q}$, under which the process $\{\tilde{W}_t\}_{t\geq 0}$ defined by

$$\tilde{W}_t = W_t + \int_0^t \gamma_s ds$$

is a standard Brownian motion. The discounted stock price, $\{\tilde{S}_t\}_{t\geq 0}$ defined as $\tilde{S}_t = S_t/B_t$, is governed by the stochastic differential equation

$$\begin{aligned} d\tilde{S}_t &= (\mu_t - r_t)\,\tilde{S}_t dt + \sigma_t \tilde{S}_t dW_t \\ &= (\mu_t - r_t - \sigma_t \gamma_t)\,\tilde{S}_t dt + \sigma_t \tilde{S}_t d\tilde{W}_t, \end{aligned}$$

and so we choose $\gamma_t = (\mu_t - r_t)/\sigma_t$. To ensure that $\{\tilde{S}_t\}_{t\geq 0}$ really is a $\mathbb{Q}$-martingale we make two further boundedness assumptions. First, in order to apply the Girsanov Theorem, we insist that

$$\mathbb{E}^{\mathbb{P}}\left[\exp\left(\int_0^T \frac{1}{2}\gamma_t^2 dt\right)\right] < \infty.$$

Second we require that $\{\tilde{S}_t\}_{t\geq 0}$ is a $\mathbb{Q}$-martingale (not just a local martingale) and so we assume a second Novikov condition:

$$\mathbb{E}^{\mathbb{Q}}\left[\exp\left(\int_0^T \frac{1}{2}\sigma_t^2 dt\right)\right] < \infty.$$

Under these extra boundedness assumptions $\{\tilde{S}_t\}_{t\geq 0}$ then is a martingale under the measure $\mathbb{Q}$ defined by

$$\left.\frac{d\mathbb{Q}}{d\mathbb{P}}\right|_{\mathcal{F}_t} = L_t^{(\gamma)} = \exp\left(-\int_0^t \gamma_s dW_s - \int_0^t \frac{1}{2}\gamma_s^2 ds\right).$$

Second step to replication

That completes the first step in our replication strategy. The second is to form the $(\mathbb{Q}, \{\mathcal{F}_t\}_{t\geq 0})$-martingale $\{M_t\}_{t\geq 0}$ given by

$$M_t = \mathbb{E}^{\mathbb{Q}}\left[B_T^{-1}C_T\,\middle|\,\mathcal{F}_t\right].$$

Replicating a   The third step is to show that our market is complete, that is any claim C_T can be
claim   replicated. First we invoke the martingale representation theorem to write

$$M_t = M_0 + \int_0^t \theta_u \, d\tilde{W}_u$$

and consequently, provided that σ_t never vanishes,

$$M_t = M_0 + \int_0^t \phi_s \, d\tilde{S}_s,$$

where $\{\phi_t\}_{t\geq 0}$ is $\{\mathcal{F}_t\}_{t\geq 0}$-predictable.

Guided by our previous work we guess that a replicating portfolio should consist
of ϕ_t units of stock and $\psi_t = M_t - \phi_t S_t$ units of cash bond at time t. In Exercise 1
it is checked that such a portfolio is self-financing. Its value at time t is

$$V_t = \phi_t S_t + \psi_t B_t = B_t M_t.$$

In particular, at time T, $V_T = B_T M_T = C_T$, and so we have a self-financing,
replicating portfolio. The usual arbitrage argument tells us that the fair value of the
claim at time t is V_t, that is the arbitrage price of the option at time t is

$$V_t = B_t \mathbb{E}^{\mathbb{Q}} \left[B_T^{-1} C_T \,\middle|\, \mathcal{F}_t \right] = \mathbb{E}^{\mathbb{Q}} \left[e^{-\int_t^T r_u \, du} C_T \,\middle|\, \mathcal{F}_t \right].$$

The   In general such an expectation must be evaluated numerically. If r_t, μ_t and σ_t
generalised   depend only on (t, S_t) then one approach to this is first to express the price as
Black–   the solution to a generalised Black–Scholes partial differential equation. This is
Scholes   achieved with the Feynman–Kac Stochastic Representation Theorem. Specifically,
equation   using Example 4.8.6, $V_t = F(t, S_t)$ where $F(t, x)$ solves

$$\frac{\partial F}{\partial t}(t, x) + \frac{1}{2}\sigma^2(t, x)x^2\frac{\partial^2 F}{\partial x^2}(t, x) + r(t, x)x\frac{\partial F}{\partial x}(t, x) - r(t, x)F(t, x) = 0,$$

subject to the terminal condition corresponding to the claim C_T, at least provided

$$\int_0^T \mathbb{E}^{\mathbb{Q}} \left[\left(\sigma(t, x)\frac{\partial F}{\partial x}(t, x) \right)^2 \right] ds < \infty.$$

For vanilla options, in the special case when r, μ and σ are functions of t alone, the
partial differential equation can be solved explicitly. As is shown in Exercise 3 the
procedure is exactly that used to solve the usual Black–Scholes equation. The price
can be found from the classical Black–Scholes price via the following simple rule:
for the value of the option at time t replace r and σ^2 by

$$\frac{1}{T-t}\int_t^T r(s)ds \quad \text{and} \quad \frac{1}{T-t}\int_t^T \sigma^2(s)ds$$

respectively.

7.2 Multiple stock models

So far we have assumed that the market consists of a riskless cash bond and a single 'risky' asset. However, the need to model whole portfolios of options or more complex equity products leads us to seek models describing several securities simultaneously. Such models must encode the *interdependence* between different security prices.

Correlated security prices

Suppose that we are modelling the evolution of n risky assets and, as ever, a single risk-free cash bond. We assume that it is not possible to exactly replicate one of the assets by a portfolio composed entirely of the others. In the most natural extension of the classical Black–Scholes model, considered individually the price of each risky asset follows a geometric Brownian motion, and interdependence of different asset prices is achieved by taking the driving Brownian motions to be correlated. Equivalently, we take a set of n independent Brownian motions and drive the asset prices by linear combinations of these; see Exercise 2. This suggests the following market model.

Multiple asset model: Our market consists of a cash bond $\{B_t\}_{0 \le t \le T}$ and n different securities with prices $\{S_t^1, S_t^2, \dots, S_t^n\}_{0 \le t \le T}$, governed by the system of stochastic differential equations

$$dB_t = rB_t dt,$$

$$dS_t^i = S_t^i \left(\sum_{j=1}^{n} \sigma_{ij}(t) dW_t^j + \mu_i(t) dt \right), \quad i = 1, 2, \dots, n, \qquad (7.3)$$

where $\{W_t^j\}_{t \ge 0}, \ j = 1, \dots, n$, are independent Brownian motions. We assume that the matrix $\sigma = (\sigma_{ij})$ is invertible.

Remarks:

1   This model is called an n-factor model as there are n sources of randomness. If there are fewer sources of randomness than stocks then there is redundancy in the model as we can replicate one of the stocks by a portfolio composed of the others. On the other hand, if we are to be able to hedge any claim in the market, then, roughly speaking, we need as many 'independent' stocks as sources of randomness. This mirrors Proposition 1.6.5.

2   Notice that the volatility of each stock in this model is really a *vector*. Since the Brownian motions $\{W_t^j\}_{t \ge 0}, \ j = 1, \dots, n$, are independent, the total volatility of the process $\{S_t^i\}_{t \ge 0}$ is $\left\{ \sqrt{\sum_{j=1}^{n} \sigma_{ij}^2(t)} \right\}_{t \ge 0}$.   □

Of course we haven't checked that this model really makes sense. That is, we need to know that the system of stochastic differential equations (7.3) has a solution. In order to verify this and to analyse such multifactor market models we need multidimensional analogues of the key results of Chapter 4.

Multifactor Itô formula

The most basic tool will be an n-factor version of the Itô formula. In the same way as we used the one-factor Itô formula to find a description (in the form of a stochastic differential equation) of models constructed as functions of Brownian motion, here we shall build new multifactor models from old. Our basic building blocks will be solutions to systems of stochastic differential equations of the form

$$dX_t^i = \mu_i(t)dt + \sum_{j=1}^{n} \sigma_{ij}(t)dW_t^j, \quad i = 1, \ldots, n, \tag{7.4}$$

where $\{W_t^j\}_{t\geq 0}$, $j = 1, \ldots, n$, are independent Brownian motions. We write $\{\mathcal{F}_t\}_{t\geq 0}$ for the σ-algebra generated by $\{W_t^j\}_{t\geq 0}$, $j = 1, \ldots, n$. Our work of Chapter 4 gives a rigorous meaning to (the integrated version of) the system (7.4) provided $\{\mu_i(t)\}_{t\geq 0}$ and $\{\sigma_{ij}(t)\}_{t\geq 0}$, $1 \leq i \leq n$, $1 \leq j \leq n$, are $\{\mathcal{F}_t\}_{t\geq 0}$-predictable processes with

$$\mathbb{E}\left[\int_0^t \left(\sum_{j=1}^{n} (\sigma_{ij}(s))^2 + |\mu_i(s)| \right) ds \right] < \infty, \quad t > 0, i = 1, \ldots, n.$$

Let us write $\{X_t\}_{t\geq 0}$ for the vector of processes $\{X_t^1, X_t^2, \ldots, X_t^n\}_{t\geq 0}$ and define a new stochastic process by $Z_t = f(t, X_t)$. Here we suppose that $f(t, x) : \mathbb{R}_+ \times \mathbb{R}^n \to \mathbb{R}$ is sufficiently smooth that we can apply Taylor's Theorem, just as in §4.3, to find the stochastic differential equation governing $\{Z_t\}_{t\geq 0}$. Writing $x = (x_1, \ldots, x_n)$, we obtain

$$dZ_t = \frac{\partial f}{\partial t}(t, X_t)dt + \sum_{i=1}^{n} \frac{\partial f}{\partial x_i}(t, X_t)dX_t^i + \frac{1}{2} \sum_{i,j=1}^{n} \frac{\partial^2 f}{\partial x_i \partial x_j}(t, X_t)dX_t^i dX_t^j + \cdots.$$

$$\tag{7.5}$$

Since the Brownian motions $\{W_t^i\}_{t\geq 0}$ are *independent* we have the multiplication table

$\times$	dW_t^i	dW_t^j	dt		
dW_t^i	dt	0	0	for $i \neq j$	(7.6)
dW_t^j	0	dt	0		
dt	0	0	0		

and this gives $dX_t^i dX_t^j = \sum_{k=1}^{n} \sigma_{ik} \sigma_{jk} dt$. The same multiplication table tells us that $dX_t^i dX_t^j dX_t^k$ is $o(dt)$ and so substituting into equation (7.5) we have provided a heuristic justification of the following result.

Theorem 7.2.1 (Multifactor Itô formula)   *Let* $\{X_t\}_{t\geq 0} = \{X_t^1, X_t^2, \ldots, X_t^n\}_{t\geq 0}$ *solve*

$$dX_t^i = \mu_i(t)dt + \sum_{j=1}^{n} \sigma_{ij}(t)dW_t^j, \qquad i = 1, 2, \ldots, n,$$

where $\{W_t^i\}_{t\geq 0}, i = 1, \ldots, n$, *are independent* $\mathbb{P}$-*Brownian motions. Further suppose that the real-valued function* $f(t,x)$ *on* $\mathbb{R}_+ \times \mathbb{R}^n$ *is continuously differentiable with respect to* t *and twice continuously differentiable in the* x-*variables. Then defining* $Z_t = f(t, X_t)$ *we have*

$$dZ_t = \frac{\partial f}{\partial t}(t, X_t)dt + \sum_{i=1}^{n} \frac{\partial f}{\partial x_i}(t, X_t)dX_t^i + \frac{1}{2}\sum_{i,j=1}^{n} \frac{\partial^2 f}{\partial x_i \partial x_j}(t, X_t)C_{ij}(t)dt$$

where $C_{ij}(t) = \sum_{k=1}^{n} \sigma_{ik}(t)\sigma_{jk}(t)$.

Remark: Notice that if we write σ for the matrix (σ_{ij}) then $C_{ij} = (\sigma\sigma^t)_{ij}$ where σ^t is the transpose of σ.   □

We can now check that there *is* a solution to the system of equations (7.3).

Example 7.2.2 (Multiple asset model)   *Let* $\{W_t^i\}_{t\geq 0}, i = 1, \ldots, n$, *be independent Brownian motions. Define* $\{S_t^1, S_t^2, \ldots, S_t^n\}_{t\geq 0}$ *by*

$$S_t^i = S_0^i \exp\left(\int_0^t \left(\mu_i(s) - \frac{1}{2}\sum_{k=1}^{n} \sigma_{ik}^2(s)\right)ds + \int_0^t \sum_{j=1}^{n} \sigma_{ij}(s)dW_s^j\right);$$

then $\{S_t^1, S_t^2, \ldots, S_t^n\}_{t\geq 0}$ *solves the system* (7.3).

Justification: Defining the processes $\{X_t^i\}_{t\geq 0}$ for $i = 1, 2, \ldots, n$ by

$$dX_t^i = \left(\mu_i(t) - \frac{1}{2}\sum_{k=1}^{n} \sigma_{ik}^2(t)\right)dt + \sum_{j=1}^{n} \sigma_{ij}(t)dW_t^j$$

we see that $S_t^i = f^i(t, X_t)$ where, writing $x = (x_1, \ldots, x_n)$, $f^i(t, x) \triangleq S_0^i e^{x_i}$. Applying Theorem 7.2.1 gives

$$
\begin{aligned}
dS_t^i &= S_0^i \exp(X_t^i)dX_t^i + \frac{1}{2}S_0^i \exp(X_t^i)C_{ii}(t)dt \\
&= S_t^i \left\{\left(\mu_i(t) - \frac{1}{2}\sum_{k=1}^{n} \sigma_{ik}^2(t)\right)dt + \sum_{j=1}^{n} \sigma_{ij}(t)dW_t^j + \frac{1}{2}\sum_{k=1}^{n} \sigma_{ik}(t)\sigma_{ik}(t)dt\right\} \\
&= S_t^i \left\{\mu_i(t)dt + \sum_{j=1}^{n} \sigma_{ij}(t)dW_t^j\right\}
\end{aligned}
$$

as required.   □

Remark: Exactly as in the single factor models, although we can write down arbitrarily complicated systems of stochastic differential equations, existence and uniqueness of solutions are far from guaranteed. If the coefficients are bounded and uniformly Lipschitz then a unique solution does exist, but such results are beyond our scope here. Instead, once again, we refer to Chung & Williams (1990) or Ikeda & Watanabe (1989).   □

Integration
by parts

We can also use the multiplication table (7.6) to write down an *n*-factor version of the integration by parts formula.

Lemma 7.2.3   *If*

$$dX_t = \mu(t, X_t)dt + \sum_{i=1}^{n} \sigma_i(t, X_t)dW_t^i$$

and

$$dY_t = v(t, Y_t)dt + \sum_{i=1}^{n} \rho_i(t, Y_t)dW_t^i$$

then

$$d(X_t Y_t) = X_t dY_t + Y_t dX_t + \sum_{i=1}^{n} \sigma_i(t, X_t)\rho_i(t, Y_t)dt.$$

Change of
measure

Pricing and hedging in the multiple stock model will follow a familiar pattern. First we find an equivalent probability measure under which *all* of the discounted stock prices $\{\tilde{S}_t^i\}_{t\geq 0}, i = 1, \ldots, n$, given by $\tilde{S}_t^i = e^{-rt}S_t^i$, are martingales. We then use a multifactor version of the Martingale Representation Theorem to construct a replicating portfolio.

Construction of the martingale measure is, of course, via a multifactor version of the Girsanov Theorem.

Theorem 7.2.4 (Multifactor Girsanov Theorem)   *Let* $\{W_t^i\}_{t\geq 0}, i = 1, \ldots, n$, *be independent Brownian motions under the measure* $\mathbb{P}$ *generating the filtration* $\{\mathcal{F}_t\}_{t\geq 0}$ *and let* $\{\theta_i(t)\}_{t\geq 0}, i = 1, \ldots, n$, *be* $\{\mathcal{F}_t\}_{t\geq 0}$-*previsible processes such that*

$$\mathbb{E}^{\mathbb{P}}\left[\exp\left(\frac{1}{2}\int_0^T \sum_{i=1}^{n}\theta_i^2(s)ds\right)\right]ds < \infty. \tag{7.7}$$

Define

$$L_t = \exp\left(-\sum_{i=1}^{n}\left(\int_0^T \theta_i(s)dW_s^i + \frac{1}{2}\int_0^T \theta_i^2(s)ds\right)\right)$$

and let $\mathbb{P}^{(L)}$ *be the probability measure defined by*

$$\left.\frac{d\mathbb{P}^{(L)}}{d\mathbb{P}}\right|_{\mathcal{F}_t} = L_t.$$

Then under $\mathbb{P}^{(L)}$ *the processes* $\{X_t^i\}_{t\geq 0}, i = 1, \ldots, n,$ *defined by*

$$X_t^i = W_t^i + \int_0^t \theta_i(s)ds$$

are all martingales.

Sketch of proof: The proof mimics that in the one-factor case. It is convenient to write $L_t = \prod_{i=1}^n L_t^i$ where

$$L_t^i = \exp\left(-\int_0^t \theta_i(s)dW_s^i - \frac{1}{2}\int_0^t \theta_i^2(s)ds\right).$$

That $\{L_t\}_{t\geq 0}$ defines a martingale follows from (7.7) and the independence of the Brownian motions $\{W_t^i\}_{t\geq 0}, i = 1, \ldots, n$.

To check that $\{X_t^i\}_{t\geq 0}$ is a (local) $\mathbb{P}^{(L)}$-martingale we find the stochastic differential equation satisfied by $\{X_t^i L_t\}_{t\geq 0}$. Since

$$dL_t^i = -\theta_i(t)L_t^i dW_t^i,$$

repeated application of our product rule gives

$$dL_t = -L_t \sum_{i=1}^n \theta_i(t)dW_t^i.$$

Moreover,

$$dX_t^i = dW_t^i + \theta_i(t)dt,$$

and so another application of our product rule gives

$$
\begin{aligned}
d(X_t^i L_t) &= X_t^i dL_t + L_t dW_t^i + L_t \theta_i(t)dt - L_t \theta_i(t)dt \\
&= -X_t^i L_t \sum_{i=1}^n \theta_i(t)dW_t^i + L_t dW_t^i.
\end{aligned}
$$

Combined with the boundedness condition (7.7), this proves that $\{X_t^i L_t\}_{t\geq 0}$ is a $\mathbb{P}$-martingale and hence $\{X_t^i\}_{t\geq 0}$ is a $\mathbb{P}^{(L)}$-martingale. $\mathbb{P}^{(L)}$ is equivalent to $\mathbb{P}$ so $\{X_t^i\}_{t\geq 0}$ has quadratic variation $[X^i]_t = t$ with $\mathbb{P}^{(L)}$-probability one and once again Lévy's characterisation of Brownian motion confirms that $\{X_t^i\}_{t\geq 0}$ is a $\mathbb{P}^{(L)}$-Brownian motion as required. □

A martingale measure

As promised we now use this to find a measure $\mathbb{Q}$, equivalent to $\mathbb{P}$, under which the discounted stock price processes $\{\tilde{S}_t^i\}_{t\geq 0}, i = 1, \ldots, n,$ are all martingales. The measure $\mathbb{Q}$ will be one of the measures $\mathbb{P}^{(L)}$ of Theorem 7.2.4. We just need to identify the appropriate drifts $\{\theta_i\}_{t\geq 0}$.

The discounted stock price $\{\tilde{S}_t^i\}_{t\geq 0}$, defined by $\tilde{S}_t^i = B_t^{-1}S_t^i$, is governed by the stochastic differential equation

$$
\begin{aligned}
d\tilde{S}_t^i &= \tilde{S}_t^i(\mu_i(t) - r)\,dt + \tilde{S}_t^i \sum_{j=1}^n \sigma_{ij}(t)dW_t^j \\
&= \tilde{S}_t^i\left(\mu_i(t) - r - \sum_{j=1}^n \theta_j(t)\sigma_{ij}(t)\right)dt + \tilde{S}_t^i \sum_{j=1}^n \sigma_{ij}(t)dX_t^j,
\end{aligned}
$$

where as in Theorem 7.2.4

$$dX_t^j = \theta_j(t)dt + dW_t^j.$$

The discounted stock price processes will (simultaneously) be (local) martingales under $\mathbb{Q} = \mathbb{P}^{(L)}$ if we can make all the drift terms vanish. That is, if we can find $\{\theta_j(t)\}_{t\geq 0}$, $j = 1, \ldots, n$, such that

$$\mu_i(t) - r - \sum_{j=1}^{n} \theta_j(t)\sigma_{ij}(t) = 0 \quad \text{for all } i = 1, \ldots, n.$$

Dropping the dependence on t in our notation and writing

$$\mu = (\mu_1, \ldots, \mu_n), \quad \theta = (\theta_1, \ldots, \theta_n), \quad \mathbf{1} = (1, \ldots, 1) \quad \text{and } \sigma = (\sigma_{ij}),$$

this becomes

$$\mu - r\mathbf{1} = \theta\sigma. \tag{7.8}$$

A solution certainly exists if the matrix σ is invertible, an assumption that we made in setting up our multiple asset model.

In order to guarantee that the discounted price processes are martingales, not just local martingales, once again we impose a Novikov condition:

$$\mathbb{E}^{\mathbb{Q}}\left[\exp\left(\int_0^t \frac{1}{2}\sum_{j=1}^{n}\sigma_{ij}^2(t)dt\right)\right] < \infty \quad \text{for each } i.$$

Replicating the claim   At this point we guess, correctly, that the value of a claim $C_T \in \mathcal{F}_T$ at time $t < T$ is its discounted expected value under the measure $\mathbb{Q}$. To prove this we show that there is a self-financing replicating portfolio and this we infer from a multifactor version of the Martingale Representation Theorem.

Theorem 7.2.5 (Multifactor Martingale Representation Theorem)   *Let*

$$\{W_t^i\}_{t\geq 0}, \quad i = 1, \ldots, n,$$

be independent $\mathbb{P}$-Brownian motions generating the filtration $\{\mathcal{F}_t\}_{t\geq 0}$. Let $\{M_t^1, \ldots, M_t^n\}_{t\geq 0}$ be given by

$$dM_t^i = \sum_{j=1}^{n}\sigma_{ij}(t)dW_t^j,$$

where

$$\mathbb{E}\left[\exp\left(\frac{1}{2}\int_0^T \sum_{j=1}^{n}\sigma_{ij}(t)^2 dt\right)\right] < \infty.$$

Suppose further that the volatility matrix $(\sigma_{ij}(t))$ is non-singular (with probability one). Then if $\{N_t\}_{t\geq 0}$ is any one-dimensional $(\mathbb{P}, \{\mathcal{F}_t\}_{t\geq 0})$-martingale there exists an n-dimensional $\{\mathcal{F}_t\}_{t\geq 0}$-previsible process $\{\phi_t\}_{t\geq 0} = \{\phi_t^1, \ldots, \phi_t^n\}_{t\geq 0}$ such that

$$N_t = N_0 + \sum_{j=1}^{n}\int_0^t \phi_s^j dM_s^j.$$

A proof of this result is beyond our scope here. It can be found, for example, in Protter (1990). Notice that the non-singularity of the matrix σ reflects our remark about non-vanishing quadratic variation after the proof of Theorem 4.6.2.

We are now in a position to verify that our guess was correct: the value of a claim in the multifactor world is its discounted expected value under the martingale measure $\mathbb{Q}$.

Let $C_T \in \mathcal{F}_T$ be a claim at time T and let $\mathbb{Q}$ be the martingale measure obtained above. We write

$$M_t = \mathbb{E}^{\mathbb{Q}} \left[B_T^{-1} C_T \,\middle|\, \mathcal{F}_t \right].$$

Since, by assumption, the matrix $\sigma = (\sigma_{ij})$ is invertible, the n-factor Martingale Representation Theorem tells us that there is an $\{\mathcal{F}_t\}_{t \geq 0}$-previsible process $\{\phi_t^1, \ldots, \phi_t^n\}_{t \geq 0}$ such that

$$M_t = M_0 + \sum_{j=1}^{n} \int_0^t \phi_s^j \, d\tilde{S}_s^j.$$

Our hedging strategy will be to hold ϕ_t^i units of the ith stock at time t for each $i = 1, \ldots, n$, and to hold ψ_t units of bond where

$$\psi_t = M_t - \sum_{j=1}^{n} \phi_t^j \tilde{S}_t^j.$$

The value of the portfolio is then $V_t = B_t M_t$, which at time T is exactly the value of the claim, and the portfolio is self-financing in that

$$dV_t = \sum_{j=1}^{n} \phi_t^j \, dS_t^j + \psi_t \, dB_t.$$

In the absence of arbitrage the value of the derivative at time t is

$$V_t = B_t \mathbb{E}^{\mathbb{Q}} \left[B_T^{-1} C_T \,\middle|\, \mathcal{F}_t \right] = e^{-r(T-t)} \mathbb{E}^{\mathbb{Q}} \left[C_T \,\middle|\, \mathcal{F}_t \right]$$

as predicted.

Remark: The multifactor market that we have constructed is complete and arbitrage-free. We have simplified the exposition by insisting that the number of sources of noise in our market is exactly matched by the number of risky tradable assets that we are modelling. More generally, we could model k risky assets driven by d sources of noise. Existence of a martingale measure corresponds to existence of a solution to (7.8). It is *uniqueness* of the martingale measure that provides us with the Martingale Representation Theorem and hence the ability to replicate any claim. For a complete arbitrage-free market we then require that $d \leq k$ and that σ has full rank. That is, the number of independent sources of randomness should exactly match the number of 'independent' risky assets trading in our market.   □

In Exercise 7 you are asked to use a delta-hedging argument to obtain this price as the solution to the multidimensional Black–Scholes equation. This partial differential equation can also be obtained directly from the expectation price and a multidimensional version of the Feynman–Kac stochastic representation. We quote the appropriate version of this useful result here.

Theorem 7.2.6 (Multidimensional Feynman–Kac stochastic representation)   *Let* $\sigma(t, x) = (\sigma_{ij}(t, x))$ *be a real symmetric* $n \times n$ *matrix,* $\Phi : \mathbb{R}^n \to \mathbb{R}$ *and* $\mu_i : \mathbb{R}_+ \times \mathbb{R}^n \to \mathbb{R}, i = 1, \ldots, n,$ *be real-valued functions and* r *be a constant. We suppose that the function* $F(t, x)$, *defined for* $(t, x) \in \mathbb{R}_+ \times \mathbb{R}^n$, *solves the boundary value problem*

$$\frac{\partial F}{\partial t}(t, x) + \sum_{i=1}^n \mu_i(t, x)\frac{\partial F}{\partial x_i}(t, x) + \frac{1}{2}\sum_{i,j=1}^n C_{ij}(t, x)\frac{\partial^2 F}{\partial x_i \partial x_j}(t, x) - rF(t, x) = 0,$$

$$F(T, x) = \Phi(x),$$

where $C_{ij}(t, x) = \sum_{k=1}^n \sigma_{ik}(t, x)\sigma_{jk}(t, x)$.
   Assume further that for each $i = 1, \ldots, n$, *the process* $\{X_t^i\}_{t \geq 0}$ *solves the stochastic differential equation*

$$dX_t^i = \mu_i(t, X_t)dt + \sum_{j=1}^n \sigma_{ij}(t, X_t)dW_t^j$$

where $X_t = \{X_t^1, \ldots, X_t^n\}$. *Finally, suppose that*

$$\int_0^T \mathbb{E}\left[\sum_{j=1}^n \left(\sigma_{ij}(s, X_s)\frac{\partial F}{\partial x_i}(s, X_s)\right)^2\right] ds < \infty, \quad i = 1, \ldots, n.$$

Then

$$F(t, x) = e^{-r(T-t)}\mathbb{E}[\Phi(X_T)| X_t = x].$$

Corollary 7.2.7   *Let* $S_t = \{S_t^1, \ldots, S_t^n\}$ *be as above and* $C_T = \Phi(S_T)$ *be a claim at time* T. *Then the price of the claim at time* $t < T$,

$$V_t = e^{-r(T-t)}\mathbb{E}^{\mathbb{Q}}[\Phi(S_T)| \mathcal{F}_t] = e^{-r(T-t)}\mathbb{E}^{\mathbb{Q}}[\Phi(S_T)| S_t = x] \triangleq F(t, x)$$

satisfies

$$\frac{\partial F}{\partial t}(t, x) + \frac{1}{2}\sum_{i,j=1}^n C_{ij}(t, x)x_i x_j\frac{\partial^2 F}{\partial x_i \partial x_j}(t, x) + r\sum_{i=1}^n x_i\frac{\partial F}{\partial x_i}(t, x) - rF(t, x) = 0,$$

$$F(T, x) = \Phi(x).$$

Proof: The process $\{S_t\}_{t \geq 0}$ is governed by

$$dS_t^i = rS_t^i dt + \sum_{j=1}^n \sigma_{ij}(t, S_t)S_t^j dX_t^j,$$

where $\{X_t^j\}_{t \geq 0}, j = 1, \ldots, n,$ are $\mathbb{Q}$-Brownian motions, so the result follows from an application of Theorem 7.2.6.   □

Numeraires   The more assets there are in our market, the more freedom we have in choosing our
'numeraire' or 'reference asset'. Usually it is chosen to be a cash bond, but in fact
it could be any of the tradable assets available. In the context of foreign exchange
we checked that we could use as reference the riskless bond in either currency and
always obtain the same value for a claim. Here we consider two numeraires in the
same market, but they may have non-zero volatility.

Suppose that our market consists of $n + 2$ tradable assets whose prices we denote
by $\{B_t^1, B_t^2, S_t^1, \ldots, S_t^n\}_{t \geq 0}$. We compare the prices obtained for a derivative by
two traders, one of whom chooses $\{B_t^1\}_{t \geq 0}$ as numeraire and the other of whom
chooses $\{B_t^2\}_{t \geq 0}$. We always assume our multidimensional geometric Brownian
motion model for the evolution of prices, but now neither of the processes $\{B_t^i\}_{t \geq 0}$
necessarily has finite variation.

If we choose $\{B_t^1\}_{t \geq 0}$ as numeraire then we first find an equivalent measure, $\mathbb{Q}^1$,
under which the asset prices discounted by $\{B_t^1\}_{t \geq 0}$, that is

$$\left\{\frac{B_t^2}{B_t^1}, \frac{S_t^1}{B_t^1}, \ldots, \frac{S_t^n}{B_t^1}\right\}_{t \geq 0},$$

are all $\mathbb{Q}^1$-martingales. The value that we obtain for a derivative with payoff C_T at
time T is then

$$V_t^1 = B_t^1 \mathbb{E}^{\mathbb{Q}^1}\left[\left.\frac{C_T}{B_T^1}\right| \mathcal{F}_t\right]$$

(see Exercise 7).

If instead we had chosen $\{B_t^2\}_{t \geq 0}$ as our numeraire then the price would have been

$$V_t^2 = B_t^2 \mathbb{E}^{\mathbb{Q}^2}\left[\left.\frac{C_T}{B_T^2}\right| \mathcal{F}_t\right]$$

where $\mathbb{Q}^2$ is an equivalent probability measure under which

$$\left\{\frac{B_t^1}{B_t^2}, \frac{S_t^1}{B_t^2}, \ldots, \frac{S_t^n}{B_t^2}\right\}_{t \geq 0}$$

are all martingales. We have not proved that such a measure $\mathbb{Q}^2$ is unique, but if
a claim can be replicated we obtain the same price for any measure $\mathbb{Q}^2$ with this
property.

Suppose that we choose $\mathbb{Q}^2$ so that its Radon–Nikodym derivative with respect to
$\mathbb{Q}^1$ is given by

$$\left.\frac{d\mathbb{Q}^2}{d\mathbb{Q}^1}\right|_{\mathcal{F}_t} = \frac{B_t^2}{B_t^1}.$$

Notice that since $\mathbb{Q}^1$ is a martingale measure for an investor choosing $\{B_t^1\}_{t \geq 0}$ as
numeraire, we know that $\{B_t^2/B_t^1\}_{t \geq 0}$ is a $\mathbb{Q}^1$-martingale. Recall that if

$$\left.\frac{d\mathbb{Q}}{d\mathbb{P}}\right|_{\mathcal{F}_t} = \zeta_t, \quad \text{for all } t > 0,$$

then, for $0 \le s \le t$,

$$\mathbb{E}^{\mathbb{Q}}\left[X_t \mid \mathcal{F}_s\right] = \mathbb{E}^{\mathbb{P}}\left[\frac{\zeta_t}{\zeta_s} X_t \mid \mathcal{F}_s\right].$$

We first apply this to check that $\{S_t^i/B_t^2\}_{t\ge 0}$ is a $\mathbb{Q}^2$-martingale for each $i = 1, \ldots, n$.

$$
\begin{aligned}
\mathbb{E}^{\mathbb{Q}^2}\left[\frac{S_t^i}{B_t^2} \mid \mathcal{F}_s\right]
&= \mathbb{E}^{\mathbb{Q}^1}\left[\frac{B_t^2}{B_t^1}\frac{B_s^1}{B_s^2}\frac{S_t^i}{B_t^2} \mid \mathcal{F}_s\right] \\
&= \mathbb{E}^{\mathbb{Q}^1}\left[\frac{B_s^1}{B_s^2}\frac{S_t^i}{B_t^1} \mid \mathcal{F}_s\right] \\
&= \frac{B_s^1}{B_s^2}\frac{S_s^i}{B_s^1} = \frac{S_s^i}{B_s^2},
\end{aligned}
$$

where the last line follows since B_s^1 and B_s^2 are $\mathcal{F}_s$-measurable and $\{S_t^i/B_t^1\}_{t\ge 0}$ is a $\mathbb{Q}^1$-martingale. In other words, $\{S_t^i/B_t^2\}_{t\ge 0}$ is a $\mathbb{Q}^2$-martingale as required. That $\{B_t^1/B_t^2\}_{t\ge 0}$ is a $\mathbb{Q}^2$-martingale follows in the same way.

The price for our derivative given that we chose $\{B_t^2\}_{t\ge 0}$ as numeraire is then

$$
\begin{aligned}
V_t^2
&= \mathbb{E}^{\mathbb{Q}^2}\left[\frac{B_t^2}{B_T^2} C_T \mid \mathcal{F}_t\right] \\
&= \mathbb{E}^{\mathbb{Q}^1}\left[\frac{B_T^2}{B_T^1}\frac{B_t^1}{B_t^2}\frac{B_t^2}{B_T^2} C_T \mid \mathcal{F}_t\right] \\
&= \mathbb{E}^{\mathbb{Q}^1}\left[\frac{B_t^1}{B_T^1} C_T \mid \mathcal{F}_t\right] = V_t^1.
\end{aligned}
$$

In other words, the choice of numeraire is unimportant – we always arrive at the same price.

Quantos

We now apply our multifactor technology in an example. We are going to price a *quanto forward contract*.

Definition 7.2.8   *A financial asset is called a* quanto product *if it is denominated in a currency other than that in which it is traded.*

A quanto forward contract is also known as a *guaranteed exchange rate forward*. It is most easily explained through an example.

Example 7.2.9   *BP, a UK company, has a Sterling denominated stock price that we denote by $\{S_t\}_{t\ge 0}$. For a dollar investor, a quanto forward contract on BP stock with maturity T has payoff $(S_T - K)$ converted into dollars according to some prearranged exchange rate. That is the payout will be $\$E(S_T - K)$ for some preagreed E, where S_T is the Sterling asset price at time T.*

As in our foreign exchange market of §5.3 we shall assume that there is a riskless cash bond in each of the dollar and Sterling markets, but now we have two random processes to model, the stock price, $\{S_t\}_{t\geq 0}$ and the exchange rate, that is the value of one pound in dollars which we denote by $\{E_t\}_{t\geq 0}$. This will then require a *two-factor* model.

> **Black–Scholes quanto model:** We write $\{B_t\}_{t\geq 0}$ for the dollar cash bond and $\{D_t\}_{t\geq 0}$ for its Sterling counterpart. Writing E_t for the dollar worth of one pound at time t and S_t for the Sterling asset price at time t, our model is
>
> $$\begin{aligned} \text{Dollar bond} \quad B_t &= e^{rt}, \\ \text{Sterling bond} \quad D_t &= e^{ut}, \\ \text{Sterling asset price} \quad S_t &= S_0 \exp\left(\nu t + \sigma_1 W_t^1\right), \\ \text{Exchange rate} \quad E_t &= E_0 \exp\left(\lambda t + \rho\sigma_2 W_t^1 + \sqrt{1-\rho^2}\sigma_2 W_t^2\right), \end{aligned}$$
>
> where $\{W_t^1\}_{t\geq 0}$ and $\{W_t^2\}_{t\geq 0}$ are independent $\mathbb{P}$-Brownian motions and r, u, ν, λ, σ_1, σ_2 and ρ are constants.

In this model the volatilities of $\{S_t\}_{t\geq 0}$ and $\{E_t\}_{t\geq 0}$ are σ_1 and σ_2 respectively and $\{W_t^1, \rho W_t^1 + \sqrt{1-\rho^2}W_t^2\}_{t\geq 0}$ is a pair of correlated Brownian motions with correlation coefficient ρ. There is no extra generality in replacing the expressions for S_t and E_t by

$$\begin{aligned} S_t &= S_0 \exp\left(\nu t + \sigma_{11}\tilde{W}_t^1 + \sigma_{12}\tilde{W}_t^2\right), \\ E_t &= E_0 \exp\left(\lambda t + \sigma_{21}\tilde{W}_t^1 + \sigma_{22}\tilde{W}_t^2\right), \end{aligned}$$

for independent Brownian motions $\{\tilde{W}_t^1, \tilde{W}_t^2\}_{t\geq 0}$.

Pricing a quanto forward contract

What is the value of K that makes the value at time zero of the quanto forward contract zero?

As in our discussion of foreign exchange, the first step is to reformulate the problem in terms of the dollar tradables. We now have three dollar tradables: the dollar worth of the Sterling bond, $E_t D_t$; the dollar worth of the stock, $E_t S_t$; and the dollar cash bond, B_t. Choosing the dollar cash bond as numeraire, we first find the stochastic differential equations governing the discounted values of the other two dollar tradables. We write $Y_t = B_t^{-1} E_t D_t$ and $Z_t = B_t^{-1} E_t S_t$. Since

$$dE_t = \left(\lambda + \frac{1}{2}\sigma_2^2\right) E_t dt + \rho\sigma_2 E_t dW_t^1 + \sqrt{1-\rho^2}\sigma_2 E_t dW_t^2,$$

application of our multifactor integration by parts formula gives

$$d(E_t D_t) = uE_t D_t dt + \left(\lambda + \frac{1}{2}\sigma_2^2\right) E_t D_t dt + \rho\sigma_2 E_t D_t dW_t^1 + \sqrt{1-\rho^2}\,\sigma_2 E_t D_t dW_t^2$$

and

$$dY_t = \left(\lambda + \frac{1}{2}\sigma_2^2 + u - r\right) Y_t dt + Y_t\left(\rho\sigma_2 dW_t^1 + \sqrt{1-\rho^2}\sigma_2 dW_t^2\right).$$

Similarly, since

$$dS_t = \left(v + \frac{1}{2}\sigma_1^2\right) S_t dt + \sigma_1 S_t dW_t^1,$$

$$
\begin{aligned}
d\left(E_t S_t\right) &= \left(v + \frac{1}{2}\sigma_1^2\right) E_t S_t dt + \sigma_1 E_t S_t dW_t^1 \\
&\quad + \left(\lambda + \frac{1}{2}\sigma_2^2\right) S_t E_t dt + \rho\sigma_2 S_t E_t dW_t^1 \\
&\quad + \sqrt{1 - \rho^2}\sigma_2 S_t E_t dW_t^2 + \rho\sigma_1\sigma_2 S_t E_t dt
\end{aligned}
$$

and so

$$
\begin{aligned}
dZ_t &= \left(v + \frac{1}{2}\sigma_1^2 + \lambda + \frac{1}{2}\sigma_2^2 + \rho\sigma_1\sigma_2 - r\right) Z_t dt \\
&\quad + (\sigma_1 + \rho\sigma_2) Z_t dW_t^1 + \sqrt{1 - \rho^2}\sigma_2 Z_t dW_t^2.
\end{aligned}
$$

Now we seek a change of measure to make these two processes martingales. Our calculations after the proof of Theorem 7.2.4 reduce this to finding θ_1, θ_2 such that

$$\lambda + \frac{1}{2}\sigma_2^2 + u - r - \theta_1\rho\sigma_2 - \theta_2\sqrt{1 - \rho^2}\sigma_2 = 0$$

and

$$v + \frac{1}{2}\sigma_1^2 + \lambda + \frac{1}{2}\sigma_2^2 + \rho\sigma_1\sigma_2 - r - \theta_1(\sigma_1 + \rho\sigma_2) - \theta_2\sqrt{1 - \rho^2}\sigma_2 = 0.$$

Solving this pair of simultaneous equations gives

$$\theta_1 = \frac{v + \frac{1}{2}\sigma_1^2 + \rho\sigma_1\sigma_2 - u}{\sigma_1}$$

and

$$\theta_2 = \frac{\lambda + \frac{1}{2}\sigma_2^2 + u - r - \rho\sigma_2\theta_1}{\sqrt{1 - \rho^2}\sigma_2}.$$

Under the martingale measure, $\mathbb{Q}$, $\{X_t^1\}_{t\geq 0}$ and $\{X_t^2\}_{t\geq 0}$ defined by $X_t^1 = W_t^1 + \theta_1 t$ and $X_t^2 = W_t^2 + \theta_2 t$ are independent Brownian motions. We have

$$S_t = S_0 \exp\left(\left(u - \rho\sigma_1\sigma_2 - \frac{1}{2}\sigma_1^2\right)t + \sigma_1 X_t^1\right).$$

In particular,

$$S_T = \exp\left(-\rho\sigma_1\sigma_2 T\right) S_0 e^{uT} \exp\left(\sigma_1 X_T^1 - \frac{1}{2}\sigma_1^2 T\right)$$

and we are finally in a position to price the forward. Since $\{X_t^1\}_{t\geq 0}$ is a $\mathbb{Q}$-Brownian motion,

$$\mathbb{E}^{\mathbb{Q}}\left[\exp\left(\sigma_1 X_T^1 - \frac{1}{2}\sigma_1^2 T\right)\right] = 1,$$

so

$$
\begin{aligned}
V_0 &= e^{-rT} \mathbb{E}\mathbb{E}^{\mathbb{Q}}\left[(S_T - K)\right] \\
&= e^{-rT} E\left(\exp\left(-\rho\sigma_1\sigma_2 T\right) S_0 e^{uT} - K\right).
\end{aligned}
$$

Writing $F = S_0 e^{uT}$ for the forward price in the Sterling market and setting $V_0 = 0$ we see that we should take

$$
K = F \exp\left(-\rho\sigma_1\sigma_2 T\right).
$$

Remark: The exchange rate is given by

$$
E_t = E_0 \exp\left(\left(r - u - \frac{1}{2}\sigma_2^2\right)t + \rho\sigma_2 X_t^1 + \sqrt{1 - \rho^2}\sigma_2 X_t^2\right).
$$

It is reassuring to observe that $\rho X_t^1 + \sqrt{1 - \rho^2} X_t^2$ is a $\mathbb{Q}$-Brownian motion with variance one so that this expression for $\{E_t\}_{t\geq 0}$ is precisely that obtained in §5.3. Notice also that the discounted stock price process $e^{-rt} S_t$ is *not* a martingale; there is an extra term, reflecting the fact that the *Sterling* price is *not* a dollar tradable. □

7.3 Asset prices with jumps

The Black–Scholes framework is highly flexible. The critical assumptions are continuous time trading and that the dynamics of the asset price are continuous. Indeed, provided this second condition is satisfied, the Black–Scholes price can be justified as an asymptotic approximation to the arbitrage price under discrete trading, as the trading interval goes to zero. But are asset prices continuous?

So far, we have always assumed that any contracts written will be honoured. In particular, if a government or company issues a bond, we have ignored the possibility that they might default on that contract at maturity. But defaults do happen. This has been dramatically illustrated in recent years by credit crises in Asia, Latin America and Russia. If a company A holds a substantial quantity of company B's debt securities, then a default by B might be expected to have the knock-on effect of a sudden drop in company A's share price. How can we incorporate these market 'shocks' into our model?

A Poisson process of jumps
By their very nature, defaults are unpredictable. If we assume that we have absolutely no information to help us predict the default times or other market shocks, then we should model them by a Poisson random variable. That is the time between shocks is exponentially distributed and the number of shocks by time t, denoted by N_t, is a Poisson random variable with parameter λt for some $\lambda > 0$. Between shocks we assume that an asset price follows our familiar geometric Brownian motion model.

A typical model for the evolution of the price of a risky asset with jumps is

$$
\frac{dS_t}{S_t} = \mu dt + \sigma dW_t - \delta dN_t, \tag{7.9}
$$

where $\{W_t\}_{t\geq 0}$ and $\{N_t\}_{t\geq 0}$ are independent. To make sense of equation (7.9) we write it in integrated form, but then we must define the stochastic integral with respect to $\{N_t\}_{t\geq 0}$. Writing τ_i for the time of the ith jump of the Poisson process, we define

$$\int_0^t f(u, S_u)dN_u = \sum_{i=1}^{N_t} f\left(\tau(i)-, S_{\tau(i)-}\right).$$

For the model (7.9), if there is a shock, then the asset price is decreased by a factor of $(1 - \delta)$. This observation tells us that the solution to (7.9) is

$$S_t = S_0 \exp\left(\left(\mu - \frac{1}{2}\sigma^2\right)t + \sigma W_t\right)(1 - \delta)^{N_t}.$$

To deal with more general models we must extend our theory of stochastic calculus to incorporate processes with jumps. As usual, the first step is to find an (extended) Itô formula.

Assumption: We assume that asset price processes are càdlàg, that is they are right continuous with left limits.

Theorem 7.3.1 (Itô's formula with jumps) *Suppose*

$$dY_t = \mu_t dt + \sigma_t dW_t + v_t dN_t$$

where, under $\mathbb{P}$, $\{W_t\}_{t\geq 0}$ is a standard Brownian motion and $\{N_t\}_{t\geq 0}$ is a Poisson process with intensity λ. If f is a twice continuously differentiable function on $\mathbb{R}$ then

$$f(Y_t) = f(Y_0) + \int_0^t f'(Y_{s-})dY_s + \frac{1}{2}\int_0^t f''(Y_{s-})\sigma_s^2 ds$$
$$- \sum_{i=1}^{N_t} f'(T_{\tau_i-})\left(Y_{\tau_i} - Y_{\tau_i-}\right) + \sum_{i=1}^{N_t}\left(f(Y_{\tau_i}) - f(Y_{\tau_i-})\right), \quad (7.10)$$

where $\{\tau_i\}$ are the times of the jumps of the Poisson process.

We don't prove this here, but heuristically it is not difficult to see that this should be the correct result. The first three terms are exactly what we'd expect if the process $\{Y_t\}_{t\geq 0}$ were continuous, but now, because of the discontinuities, we must distinguish Y_{s-} from Y_s. In between jumps of $\{N_t\}_{t\geq 0}$, precisely this equation should apply, but we must compensate for changes at jump times. In the first three terms we have included a term of the form $\sum_{i=1}^{N_t} f'(Y_{\tau_i-})\left(Y_{\tau_i} - Y_{\tau_i-}\right)$ and the first sum in equation (7.10) corrects for this. Since N_t is finite, we do not have to correct the term involving f''. Now we add in the *actual* contribution from the jump times and this is the second sum.

Compensation   As usual a key rôle will be played by martingales. Evidently a Poisson process, $\{N_t\}_{t\geq 0}$ with intensity λ under $\mathbb{P}$ is not a $\mathbb{P}$-martingale – it is monotone increasing. But we can write it as a martingale plus a drift. In Exercise 13 it is shown that the process $\{M_t\}_{t\geq 0}$ defined by $M_t = N_t - \lambda t$ is a $\mathbb{P}$-martingale.

More generally we can consider time-inhomogeneous Poisson processes. For such processes the intensity $\{\lambda_t\}_{t\geq 0}$ is a function of time. The probability of a jump in the time interval $[t, t + \delta t)$ is $\lambda_t \delta t + o(\delta t)$. Thus, for example, the probability that there is no jump in the interval $[s, t]$ is $\exp\left(-\int_s^t \lambda_u du\right)$. The corresponding *Poisson martingale* is $M_t = N_t - \int_0^t \lambda_s ds$. The process $\{\Lambda_t\}_{t\geq 0}$ given by $\Lambda_t = \int_0^t \lambda_s ds$ is the *compensator* of $\{N_t\}_{t\geq 0}$.

In Exercise 14 it is shown that just as integration with respect to Brownian martingales gives rise to (local) martingales, so integration with respect to Poisson martingales gives rise to martingales.

Poisson   **Example 7.3.2**   *Let $\{N_t\}_{t\geq 0}$ be a Poisson process with intensity $\{\lambda_t\}_{t\geq 0}$ under $\mathbb{P}$*
exponential   *where for each $t > 0$, $\int_0^t \lambda_s ds < \infty$. For a given bounded deterministic function*
martingales   *$\{\alpha_t\}_{t\geq 0}$, let*

$$L_t = \exp\left(\int_0^t \alpha_s dM_s + \int_0^t \left(1 + \alpha_s - e^{\alpha_s}\right) \lambda_s ds\right) \qquad (7.11)$$

where $dM_s = dN_s - \lambda_s ds$. Find the stochastic differential equation satisfied by $\{L_t\}_{t\geq 0}$ and deduce that $\{L_t\}_{t\geq 0}$ is a $\mathbb{P}$-martingale.

Solution: First write

$$Z_t = \int_0^t \alpha_s dM_s + \int_0^t \left(1 + \alpha_s - e^{\alpha_s}\right) \lambda_s ds$$

so that $L_t = e^{Z_t}$. Then

$$dZ_t = \alpha_t dN_t - \alpha_t \lambda_t dt + \left(1 + \alpha_t - e^{\alpha_t}\right) \lambda_t dt$$

and by our generalised Itô formula

$$dL_t = L_{t-} dZ_t + \left(-e^{Z_{t-}} \alpha_t + e^{Z_{t-}+\alpha_t} - e^{Z_{t-}}\right) dN_t,$$

where we have used the fact that if a jump in $\{Z_t\}_{t\geq 0}$ takes place at time t, then that jump is of size α_{t-}. Substituting and rearranging give

$$\begin{aligned} dL_t &= L_{t-}\alpha_t dM_t + L_{t-}\left(1 + \alpha_t - e^{\alpha_t}\right) \lambda_t dt - L_{t-}\left(1 + \alpha_t - e^{\alpha_t}\right) dN_t \\ &= L_{t-}\left(e^{\alpha_t} - 1\right) dM_t. \end{aligned}$$

By Exercise 14, $\{L_t\}_{t\geq 0}$ is a $\mathbb{P}$-martingale.   □

Definition 7.3.3   *Processes of the form of $\{L_t\}_{t\geq 0}$ defined by (7.11) will be called* Poisson exponential martingales.

Our Poisson exponential martingales and Brownian exponential martingales are examples of Doléans–Dade exponentials.

Definition 7.3.4   *For a semimartingale $\{X_t\}_{t\geq 0}$ with $X_0 = 0$, the* Doléans–Dade *exponential of $\{X_t\}_{t\geq 0}$ is the unique semimartingale solution $\{Z_t\}_{t\geq 0}$ to*

$$Z_t = 1 + \int_0^t Z_{s-} dX_s.$$

Change of measure

In the same way as we used Brownian exponential martingales to change measure and thus 'transform drift' in the continuous world, so we shall combine Brownian and Poisson exponential martingales in our discontinuous asset pricing models. A change of drift for a Poisson martingale will correspond to a change of intensity for the Poisson process $\{N_t\}_{t\geq 0}$. More precisely, we have the following version of the Girsanov Theorem.

Theorem 7.3.5 (Girsanov Theorem for asset prices with jumps)   *Let $\{W_t\}_{t\geq 0}$ be a standard $\mathbb{P}$-Brownian motion and $\{N_t\}_{t\geq 0}$ a (possibly time-inhomogeneous) Poisson process with intensity $\{\lambda_t\}_{t\geq 0}$ under $\mathbb{P}$. That is*

$$M_t = N_t - \int_0^t \lambda_u du$$

is a $\mathbb{P}$-martingale. We write $\mathcal{F}_t$ for the σ-field generated by $\mathcal{F}_t^W \cup \mathcal{F}_t^N$. Suppose that $\{\theta_t\}_{t\geq 0}$ and $\{\phi_t\}_{t\geq 0}$ are $\{\mathcal{F}_t\}_{t\geq 0}$-previsible processes with ϕ_t positive for each t, such that

$$\int_0^t \|\theta_s\|^2 ds < \infty \quad and \quad \int_0^t \phi_s \lambda_s ds < \infty.$$

Then under the measure $\mathbb{Q}$ whose Radon–Nikodym derivative with respect to $\mathbb{P}$ is given by

$$\left.\frac{d\mathbb{Q}}{d\mathbb{P}}\right|_{\mathcal{F}_t} = L_t$$

where $L_0 = 1$ and

$$\frac{dL_t}{L_{t-}} = \theta_t dW_t - (1 - \phi_t) dM_t,$$

the process $\{X_t\}_{t\geq 0}$ defined by $X_t = W_t - \int_0^t \theta_s ds$ is a Brownian motion and $\{N_t\}_{t\geq 0}$ has intensity $\{\phi_t \lambda_t\}_{t\geq 0}$.

In Exercise 16 it is shown that $\{L_t\}_{t\geq 0}$ is actually the product of a Brownian exponential martingale and a Poisson exponential martingale.

The proof of Theorem 7.3.5 is once again beyond our scope, but to check that the processes $\{X_t\}_{t\geq 0}$ and $\{N_t - \int_0^t \phi_s \lambda_s ds\}_{t\geq 0}$ are both local martingales under $\mathbb{Q}$ is an exercise based on the Itô formula.

Heuristics: An informal justification of the result is based on the extended multiplication table:

×	dW_t	dN_t	dt
dW_t	dt	0	0
dN_t	0	dN_t	0
dt	0	0	0

Thus, for example,

$$
\begin{aligned}
d\left(L_t\left(N_t - \int_0^t \phi_s \lambda_s ds\right)\right) &= \left(N_t - \int_0^t \phi_s \lambda_s ds\right) dL_t + L_t (dN_t - \phi_t \lambda_t dt) \\
&\quad - L_t(1 - \phi_t)(dN_t)^2 \\
&= \left(N_t - \int_0^t \phi_s \lambda_s ds\right) dL_t + L_t (dM_t + \lambda_t dt) \\
&\quad - L_t \phi_t \lambda_t dt - L_t(1 - \phi_t)(dM_t + \lambda_t dt) \\
&= \left(N_t - \int_0^t \phi_s \lambda_s ds\right) dL_t + L_t \phi_t dM_t.
\end{aligned}
$$

Since $\{M_t\}_{t\geq 0}$ and $\{L_t\}_{t\geq 0}$ are $\mathbb{P}$-martingales, subject to appropriate boundedness assumptions, $\left\{L_t\left(N_t - \int_0^t \phi_s \lambda_s ds\right)\right\}_{t\geq 0}$ should be a $\mathbb{P}$-martingale and consequently $\left\{\left(N_t - \int_0^t \phi_s \lambda_s ds\right)\right\}_{t\geq 0}$ should be a $\mathbb{Q}$-martingale.   □

Our instinct is to use the extended Girsanov Theorem to find an equivalent probability measure under which the discounted asset price is a martingale.

Suppose then that
$$
\frac{dS_t}{S_t} = \mu dt + \sigma dW_t - \delta dN_t.
$$

Evidently the discounted asset price satisfies
$$
\frac{d\tilde{S}_t}{\tilde{S}_t} = (\mu - r) dt + \sigma dW_t - \delta dN_t.
$$

But now we see that there are *many* choices of $\{\theta_t\}_{t\geq 0}$ and $\{\phi_t\}_{t\geq 0}$ in Theorem 7.3.5 that lead to a martingale measure. The difficulty of course is that our market is not *complete*, so that although for any replicable claim we can use any of the martingale measures and arrive at the same answer, there are claims that cannot be hedged. There are two independent sources of risk, the Brownian motion and the Poisson point process, and so if we are to be able to hedge arbitrary claims $C_T \in \mathcal{F}_T$, we need two tradable risky assets subject to the same two noises.

Market price of risk   So if there *are* enough assets available to hedge claims, can we find a measure under which once discounted they are *all* martingales? Remember that otherwise there will be arbitrage opportunities in our market.

If the asset price has no jumps, we can write
$$
\begin{aligned}
\frac{dS_t}{S_t} &= \mu dt + \sigma dW_t \\
&= (r + \gamma\sigma) dt + \sigma dW_t,
\end{aligned}
$$

where $\gamma = (\mu - r)/\sigma$ is the market price of risk. We saw in Chapter 5 that in the absence of arbitrage (so when there *is* an equivalent martingale measure for our market), γ will be the same for *all* assets driven by $\{W_t\}_{t\geq 0}$.

If the asset price has jumps, then investors will expect to be compensated for the additional risk associated with the possibility of downward jumps, even if we have 'compensated' the jumps (replaced dN_t by dM_t) so that their mean is zero. The price of such an asset is governed by

$$\frac{dS_t}{S_t} = \mu dt + \sigma dW_t + v dM_t$$
$$= (r + \gamma \sigma + \eta \lambda v) \, dt + \sigma dW_t + v dM_t$$

where v measures the sensitivity of the asset price to the market shock and η is the excess rate of return per unit of jump risk. Again if there is to be a martingale measure under which *all* the discounted asset prices are martingales, then σ and η should be the same for all assets whose prices are driven by $\{W_t\}_{t \geq 0}$ and $\{N_t\}_{t \geq 0}$. The martingale measure, $\mathbb{Q}$, will then be the measure $\mathbb{Q}$ of Theorem 7.3.5 under which

$$W_t + \int_0^t \frac{\mu - r}{\sigma} ds \quad \text{and} \quad M_t - \int_0^t \eta \lambda ds$$

are martingales. That is we take $\theta = \gamma$ and $\phi = -\eta$.

Multiple noises

The same ideas can be extended to assets driven by larger numbers of independent noises. For example, we might have n assets with dynamics

$$\frac{dS_t^i}{S_t^i} = \mu_i dt + \sum_{\alpha=1}^{n} \sigma_{i\alpha} dW_t^\alpha + \sum_{\beta=1}^{m} v_{i\beta} dM_t^\beta$$

where, under $\mathbb{P}$, $\{W_t^\alpha\}_{t \geq 0}$, $\alpha = 1, \ldots, n$, are independent Brownian motions and $\{M_t^\beta\}_{t \geq 0}$, $\beta = 1, \ldots, m$, are independent Poisson martingales.

There will be an equivalent martingale measure under which *all* the discounted asset prices are martingales if we can associate a unique market price of risk with each source of noise. In this case we can write

$$\mu_i = r + \sum_{\alpha=1}^{n} \gamma_\alpha \sigma_{i\alpha} + \sum_{\beta=1}^{m} \eta_\beta \lambda_\beta v_{i\beta}.$$

All discounted asset prices will be martingales under the measure $\mathbb{Q}$ for which

$$\tilde{W}_t^\alpha = W_t^\alpha + \gamma_\alpha t$$

is a martingale for each α and

$$\tilde{M}_t^\beta = M_t^\beta + \eta_\beta \lambda_\beta t$$

is a martingale for each β.

As always it is *replication* that drives the theory. Note that in order to be able to hedge arbitrary $C_T \in \mathcal{F}_T$ we'll require $n + m$ 'independent' tradable risky assets driven by these sources of noise. With fewer assets at our disposal there will be claims C_T that we cannot hedge.

All this is little changed if we take the coefficients μ, σ, λ to be adapted to the filtration generated by $\{W_t^i\}_{t \geq 0}$, $i = 1, \ldots, n$; see Exercise 15. Since we are not introducing any extra sources of noise, the same number of assets will be needed for market completeness. These ideas form the basis of Jarrow–Madan theory.

7.4   Model error

Even in the absence of jumps (or between jumps) we have given only a very vague justification for the Samuelson model

$$dS_t = \mu S_t dt + \sigma S_t dW_t. \tag{7.12}$$

Moreover, although we have shown that under this model the pricing and hedging of derivatives are dictated by the single parameter σ, we have said nothing about how actually to estimate this number from market data. So what is market practice?

Implied
volatility

Vanilla options are generally traded on exchanges, so if a trader wants to know the price of, say, a European call option, then she can read it from her trading screen. However, for an over-the-counter derivative, the price is not quoted on an exchange and so one needs a pricing model. The normal practice is to build a Black–Scholes model and then *calibrate* it to the market – that is estimate σ from the market. But it is *not* usual to estimate σ directly from data for the stock price. Instead one uses the quoted price for exchange-traded options written on the same stock. The procedure is simple: for given strike price and maturity, we can think of the Black–Scholes pricing formula for a European option as a mapping from volatility, σ, to price V. In Exercise 17, it is shown that for vanilla options this mapping is strictly monotone and so can be inverted to infer σ from the price. In other words, given the option price one can recover the corresponding value of σ in the Black–Scholes formula. This number is the so-called *implied volatility*.

If the markets really did follow our Black–Scholes model, then this procedure would give the same value of σ, irrespective of the strike price and maturity of the exchange-traded option chosen. Sadly, this is far from what we observe in reality: not only is there dependence on the strike price for a fixed maturity, giving rise to the famous volatility smile, but also implied volatility tends to increase with time to maturity (Figure 7.1). Market practice is to choose as volatility parameter for pricing an over-the-counter option the implied volatility obtained from 'comparable' exchange-traded options.

Hedging
error

This procedure can be expected to lead to a *consistent* price for exchange-traded and over-the-counter options and model error is not a serious problem. The difficulties arise in *hedging*. Even for exchange-traded options a model is required to determine the replicating portfolio. We follow Davis (2001).

Suppose that the true stock price process follows

$$dS_t = \alpha_t S_t dS_t + \beta_t S_t dW_t$$

where $\{\alpha_t\}_{t \geq 0}$ and $\{\beta_t\}_{t \geq 0}$ are $\{\mathcal{F}_t\}_{t \geq 0}$-adapted processes, but we price *and hedge* an option with payoff $\Phi(S_T)$ at time T as though $\{S_t\}_{t \geq 0}$ followed equation (7.12) for some parameter σ.

Our estimate for the value of the option at time $t < T$ will be $V(t, S_t)$ where

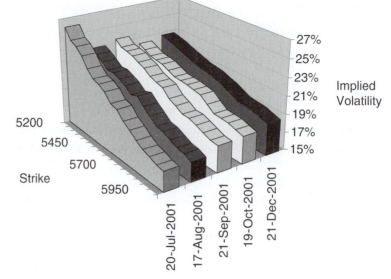

Figure 7.1   Implied volatility as a function of strike price and maturity for European call options based on the FTSE stock index.

$V(t, x)$ satisfies the Black–Scholes partial differential equation

$$\frac{\partial V}{\partial t}(t, x) + rx\frac{\partial V}{\partial x}(t, x) + \frac{1}{2}\sigma^2 x^2\frac{\partial^2 V}{\partial x^2}(t, x) - rV(t, x) = 0,$$
$$V(T, x) = \Phi(x).$$

Our hedging portfolio consists at time t of $\phi_t = \frac{\partial V}{\partial x}(t, S_t)$ units of stock and cash bonds with total value $\psi_t e^{rt} \triangleq V(t, S_t) - \phi_t S_t$.

Our first worry is that because of model misspecification, the portfolio is not self-financing. So what is the cost of following such a strategy? Since the cost of purchasing the 'hedging' portfolio at time t is $V(t, S_t)$, the incremental cost of the strategy over an infinitesimal time interval $[t, t + \delta t)$ is

$$\frac{\partial V}{\partial x}(t, S_t)\big(S_{t+\delta t} - S_t\big) + \left(V(t, S_t) - \frac{\partial V}{\partial x}(t, S_t)S_t\right)(e^{r\delta t} - 1)$$
$$- V(t + \delta t, S_{t+\delta t}) + V(t, S_t).$$

In other words, writing Z_t for our net position at time t, we have

$$dZ_t = \frac{\partial V}{\partial x}(t, S_t)dS_t + \left(V(t, S_t) - \frac{\partial V}{\partial x}(t, S_t)S_t\right)rdt - dV(t, S_t).$$

Since $V(t, x)$ solves the Black–Scholes partial differential equation, applying Itô's formula gives

$$
\begin{aligned}
dZ_t &= \frac{\partial V}{\partial x}(t, S_t)dS_t + \left(V(t, S_t) - \frac{\partial V}{\partial x}(t, S_t)S_t\right)rdt \\
&\quad - \frac{\partial V}{\partial t}(t, S_t)dt - \frac{\partial V}{\partial x}(t, S_t)dS_t - \frac{1}{2}\frac{\partial^2 V}{\partial x^2}(t, S_t)\beta_t^2 S_t^2 dt \\
&= \frac{1}{2}S_t^2\frac{\partial^2 V}{\partial x^2}\left(\sigma^2 - \beta_t^2\right)dt.
\end{aligned}
$$

Irrespective of the model, $V(T, S_T) = \Phi(S_T)$ precisely matches the claim against us at time T, so our net position at time T (having honoured the claim $\Phi(S_T)$ against us) is

$$
Z_T = \int_0^T \frac{1}{2}S_t^2\frac{\partial^2 V}{\partial x^2}(t, S_t)\left(\sigma^2 - \beta_t^2\right)dt.
$$

For European call and put options $\frac{\partial^2 V}{\partial x^2} > 0$ (see Exercise 18) and so if $\sigma^2 > \beta_t^2$ for all $t \in [0, T]$ our hedging strategy makes a profit. This means that regardless of the price dynamics, we make a profit if the parameter σ in our Black–Scholes model dominates the true diffusion coefficient β. This is key to successful hedging. Our calculation won't work if the price process has jumps, although by choosing σ large enough one can still arrange for Z_T to have positive expectation.

   The choice of σ is still a tricky matter. If we are too cautious no one will buy the option, too optimistic and we are exposed to the risk associated with changes in volatility and we should try to hedge that risk. Such hedging is known as *vega hedging*, the Greek *vega* of an option being the sensitivity of its Black–Scholes price to changes in σ. The idea is the same as that of delta hedging (Exercise 5 of Chapter 5). For example, if we buy an over-the-counter option for which $\frac{\partial V}{\partial \sigma} = v$, then we also sell a number v/v' of a comparable exchange traded option whose value is V' and for which $\frac{\partial V'}{\partial \sigma} = v'$. The resulting portfolio is said to be *vega-neutral*.

Stochastic volatility and implied volatility

Since we cannot observe the volatility directly, it is natural to try to model it as a random process. A huge amount of effort has gone into developing so-called *stochastic volatility models*. Fat-tailed returns distributions observed in data can be modelled in this framework and sometimes 'jumps' in the asset price can be best modelled by jumps in the volatility. For example if jumps occur according to a Poisson process with constant rate and at the time, τ, of a jump, $S_\tau/S_{\tau-}$ has a lognormal distribution, then the distribution of S_t will be lognormal but with variance parameter given by a multiple of a Poisson random variable (Exercise 19). Stochastic volatility can also be used to model the 'smile' in the implied volatility curve and we end this chapter by finding the correspondence between the choice of a stochastic volatility model and of an implied volatility model. Once again we follow Davis (2001). A typical stochastic volatility model takes the form

$$
\begin{aligned}
dS_t &= \mu S_t dt + \sigma_t S_t dW_t^1, \\
d\sigma_t &= a(S_t, \sigma_t)dt + b(S_t, \sigma_t)\left(\rho dW_t^1 + \sqrt{1 - \rho^2}dW_t^2\right),
\end{aligned}
$$

where $\{W_t^1\}_{t\geq0}$, $\{W_t^2\}_{t\geq0}$ are independent $\mathbb{P}$-Brownian motions, ρ is a constant in $(0, 1)$ and the coefficients $a(x, \sigma)$ and $b(x, \sigma)$ define the volatility model.

As usual we'd like to find a martingale measure. If $\mathbb{Q}$ is equivalent to $\mathbb{P}$, then its Radon–Nikodym derivative with respect to $\mathbb{P}$ takes the form

$$\left.\frac{d\mathbb{Q}}{d\mathbb{P}}\right|_{\mathcal{F}_t} = \exp\left(-\int_0^t \hat{\theta}_s dW_s^1 - \frac{1}{2}\int_0^t \hat{\theta}_s^2 ds - \int_0^t \theta_s dW_s^2 - \frac{1}{2}\int_0^t \theta_s^2 ds\right)$$

for some integrands $\{\hat{\theta}_t\}_{t\geq0}$ and $\{\theta_t\}_{t\geq0}$. In order for the discounted asset price $\{\tilde{S}_t\}_{t\geq0}$ to be a $\mathbb{Q}$-martingale, we choose

$$\hat{\theta}_t = \frac{\mu - r}{\sigma_t}.$$

The choice of $\{\theta_t\}_{t\geq0}$ however is arbitrary as $\{\sigma_t\}_{t\geq0}$ is not a tradable and so no arbitrage argument can be brought to bear to dictate its drift. Under $\mathbb{Q}$,

$$X_t^1 = W_t^1 + \int_0^t \hat{\theta}_s ds$$

and

$$X_t^2 = W_t^2 + \int_0^t \theta_s ds$$

are independent Brownian motions. The dynamics of $\{S_t\}_{t\geq0}$ and $\{\sigma_t\}_{t\geq0}$ are then most conveniently written as

$$dS_t = rS_t dt + \sigma_t S_t dX_t^1$$

and

$$d\sigma_t = \tilde{a}(S_t, \sigma_t)dt + b(S_t, \sigma_t)\left(\rho dX_t^1 + \sqrt{1-\rho^2}dX_t^2\right)$$

where

$$\tilde{a}(S_t, \sigma_t) = a(S_t, \sigma_t) - b(S_t, \sigma_t)\left(\rho\hat{\theta}_t + \sqrt{1-\rho^2}\theta_t\right).$$

We now *introduce* a second tradable asset. Suppose that we have an option written on $\{S_t\}_{t\geq0}$ whose exercise value at time T is $\Phi(S_T)$. We *define* its value at times $t < T$ to be the discounted value of $\Phi(S_T)$ under the measure $\mathbb{Q}$. That is

$$V(t, S_t, \sigma_t) = \mathbb{E}^{\mathbb{Q}}\left[e^{-r(T-t)}\Phi(S_T)\,\middle|\,\mathcal{F}_t\right].$$

Our multidimensional Feynman–Kac Stochastic Representation Theorem (combined with the usual product rule) tells us that the function $V(t, x, \sigma)$ solves the partial differential equation

$$\frac{\partial V}{\partial t}(t, x, \sigma) + rx\frac{\partial V}{\partial x}(t, x, \sigma) + \tilde{a}(t, x, \sigma)\frac{\partial V}{\partial \sigma}(t, x, \sigma) + \frac{1}{2}\sigma^2 x^2\frac{\partial^2 V}{\partial x^2}(t, x, \sigma)$$

$$+ \frac{1}{2}b(t, x, \sigma)^2\frac{\partial^2 V}{\partial \sigma^2}(t, x, \sigma) + \rho\sigma xb(t, x, \sigma)\frac{\partial^2 V}{\partial x\partial \sigma}(t, x, \sigma) - rV(t, x, \sigma) = 0.$$

Writing $Y_t = V(t, S_t, \sigma_t)$ and suppressing the dependence of V, $\tilde{a}$ and b on (t, S_t, σ_t) in our notation, an application of Itô's formula tells us that

$$
\begin{aligned}
dY_t &= \frac{\partial V}{\partial t}dt + \frac{\partial V}{\partial x}dS_t + \frac{\partial V}{\partial \sigma}d\sigma_t + \frac{1}{2}\frac{\partial^2 V}{\partial x^2}\sigma_t^2 S_t^2 dt \\
&\quad + \frac{\partial^2 V}{\partial x \partial \sigma}\rho b \sigma_t S_t dt + \frac{1}{2}\frac{\partial^2 V}{\partial \sigma^2}b^2 dt \\
&= \left(rV - rS_t\frac{\partial V}{\partial x} - \tilde{a}\frac{\partial V}{\partial \sigma} - \frac{1}{2}\sigma_t^2 S_t^2\frac{\partial^2 V}{\partial x^2} - \frac{1}{2}b^2\frac{\partial^2 V}{\partial \sigma^2} - \rho\sigma_t S_t b\frac{\partial^2 V}{\partial x \partial \sigma}\right)dt \\
&\quad + rS_t\frac{\partial V}{\partial x}dt + \sigma_t S_t\frac{\partial V}{\partial x}dX_t^1 + \tilde{a}\frac{\partial V}{\partial \sigma}dt + b\rho\frac{\partial V}{\partial \sigma}dX_t^1 + b\sqrt{1-\rho^2}\frac{\partial V}{\partial \sigma}dX_t^2 \\
&\quad + \frac{1}{2}\sigma_t^2 S_t^2\frac{\partial^2 V}{\partial x^2}dt + \rho b\sigma_t S_t\frac{\partial^2 V}{\partial x \partial \sigma}dt + \frac{1}{2}b^2\frac{\partial^2 V}{\partial \sigma^2}dt \\
&= rY_t dt + \sigma_t S_t\frac{\partial V}{\partial x}dX_t^1 + b\rho\frac{\partial V}{\partial \sigma}dX_t^1 + b\sqrt{1-\rho^2}\frac{\partial V}{\partial \sigma}dX_t^2.
\end{aligned}
$$

If the mapping $\sigma \mapsto y = V(t, x, \sigma)$ is invertible so that $\sigma = D(t, x, y)$ for some nice function D, then

$$
dY_t = rY_t dt + c(t, S_t, Y_t)dX_t^1 + d(t, S_t, Y_t)dX_t^2
$$

for some functions c and d.

We have now created a complete market model with tradables $\{S_t\}_{t \geq 0}$ and $\{Y_t\}_{t \geq 0}$ for which $\mathbb{Q}$ is the unique martingale measure. Of course, we have actually created one such market for each choice of $\{\theta_t\}_{t \geq 0}$ and it is the choice of $\{\theta_t\}_{t \geq 0}$ that specifies the functions c and d and it is precisely these functions that tell us how to hedge.

So what model for implied volatility corresponds to this stochastic volatility model? The implied volatility, $\hat{\sigma}(t)$, will be such that Y_t is the Black–Scholes price evaluated at (t, S_t) if the volatility in equation (7.12) is taken to be $\hat{\sigma}(t)$. In this way each choice of $\{\theta_t\}_{t \geq 0}$, or equivalently model for $\{Y_t\}_{t \geq 0}$, provides a model for the implied volatility.

There is a huge literature on stochastic volatility. A good starting point is Fouque, Papanicolau and Sircar (2000).

Exercises

1   Check that the replicating portfolio defined in §7.1 is self-financing.

2   Suppose that $\{W_t^1\}_{t \geq 0}$ and $\{W_t^2\}_{t \geq 0}$ are independent Brownian motions under $\mathbb{P}$ and let ρ be a constant with $0 < \rho < 1$. Find coefficients $\{\alpha_{ij}\}_{i,j=1,2}$ such that

$$
\tilde{W}_t^1 = \alpha_{11}W_t^1 + \alpha_{12}W_t^2
$$

and

$$
\tilde{W}_t^2 = \alpha_{21}W_t^1 + \alpha_{22}W_t^2
$$

define two standard Brownian motions under $\mathbb{P}$ with $\mathbb{E}\left[\tilde{W}_t^1 \tilde{W}_t^2\right] = \rho t$. Is your solution unique?

3 Suppose that $F(t, x)$ solves the time-inhomogeneous Black–Scholes partial differential equation

$$\frac{\partial F}{\partial t}(t, x) + \frac{1}{2}\sigma^2(t)x^2\frac{\partial^2 F}{\partial x^2}(t, x) + r(t)x\frac{\partial F}{\partial x}(t, x) - r(t)F(t, x) = 0, \qquad (7.13)$$

   subject to the boundary conditions appropriate to pricing a European call option.
   Substitute

$$y = xe^{\alpha(t)}, \qquad v = Fe^{\beta(t)}, \qquad \tau = \gamma(t)$$

   and choose $\alpha(t)$ and $\beta(t)$ to eliminate the coefficients of v and $\frac{\partial v}{\partial y}$ in the resulting
   equation and $\gamma(t)$ to remove the remaining time dependence so that the equation
   becomes

$$\frac{\partial v}{\partial \tau}(\tau, y) = \frac{1}{2}y^2\frac{\partial^2 v}{\partial y^2}(\tau, y).$$

   Notice that the coefficients in this equation are independent of time and there is no
   reference to r or σ. Deduce that the solution to equation (7.13) can be obtained by
   making appropriate substitutions in the classical Black–Scholes formula.

4 Let $\{W_t^i\}_{t\geq 0}, i = 1, \ldots, n$, be independent Brownian motions. Show that $\{R_t\}_{t\geq 0}$
   defined by

$$R_t = \sqrt{\sum_{i=1}^{n}(W_t^i)^2}$$

   satisfies a stochastic differential equation. The process $\{R_t\}_{t\geq 0}$ is the radial part of
   Brownian motion in $\mathbb{R}^n$ and is known as the *n-dimensional Bessel process*.

5 Recall that we define two-dimensional Brownian motion, $\{X_t\}_{t\geq 0}$, by $X_t =$
   (W_t^1, W_t^2), where $\{W_t^1\}_{t\geq 0}$ and $\{W_t^2\}_{t\geq 0}$ are independent (one-dimensional) standard
   Brownian motions. Find the Kolmogorov backward equation for $\{X_t\}_{t\geq 0}$.
   Repeat your calculation if $\{W_t^1\}_{t\geq 0}$ and $\{W_t^2\}_{t\geq 0}$ are replaced by *correlated* Brown-
   ian motions, $\{\tilde{W}_t^1\}_{t\geq 0}$ and $\{\tilde{W}_t^1\}_{t\geq 0}$ with $\mathbb{E}[d\tilde{W}_t^1 d\tilde{W}_t^2] = \rho dt$ for some $-1 < \rho < 1$.

6 Use a delta-hedging argument to obtain the result of Corollary 7.2.7.

7 Repeat the Black–Scholes analysis of §7.2 in the case when the chosen numeraire,
   $\{B_t\}_{t\geq 0}$, has non-zero volatility and check that the fair price of a derivative with
   payoff C_T at time T is once again

$$V_t = B_t\mathbb{E}^{\mathbb{Q}}\left[\frac{C_T}{B_T}\bigg|\mathcal{F}_t\right]$$

   for a suitable choice of $\mathbb{Q}$ (which you should specify).

8 Two traders, operating in the same complete arbitrage-free Black–Scholes market of
   §7.2, sell identical options, but make different choices of numeraire. How will their
   hedging strategies differ?

9 Find a portfolio that replicates the quanto forward contract of Example 7.2.9.

10 A *quanto digital contract* written on the BP stock of Example 7.2.9 pays $1 at time T if the BP Sterling stock price, S_T, is larger than K. Assuming the Black–Scholes quanto model of §7.2, find the time zero price of such an option and the replicating portfolio.

11 A *quanto call option* written on the BP stock of Example 7.2.9 is worth $E(S_T - K)_+$ dollars at time T, where S_T is the *Sterling* stock price. Assuming the Black–Scholes quanto model of §7.2, find the time zero price of the option and the replicating portfolio.

12 *Asian options* Suppose that our market, consisting of a riskless cash bond, $\{B_t\}_{t\geq 0}$, and a single risky asset with price $\{S_t\}_{t\geq 0}$, is governed by

$$dB_t = rB_t dt, \quad B_0 = 1$$

and

$$dS_t = \mu S_t dt + \sigma S_t dW_t,$$

where $\{W_t\}_{t\geq 0}$ is a $\mathbb{P}$-Brownian motion.
An option is written with payoff $C_T = \Phi(S_T, Z_T)$ at time T where

$$Z_t = \int_0^t g(u, S_u) du$$

for some (deterministic) real-valued function g on $\mathbb{R}_+ \times \mathbb{R}$.
From our general theory we know that the value of such an option at time t satisfies

$$V_t = e^{-r(T-t)}\mathbb{E}^{\mathbb{Q}}\left[\Phi(S_T, Z_T)\mid \mathcal{F}_t\right]$$

where $\mathbb{Q}$ is the measure under which $\{S_t/B_t\}_{t\geq 0}$ is a martingale.
Show that $V_t = F(t, S_t, Z_t)$ where the real-valued function $F(t, x, z)$ on $\mathbb{R}_+ \times \mathbb{R} \times \mathbb{R}$ solves

$$\frac{\partial F}{\partial t} + rx\frac{\partial F}{\partial x} + \frac{1}{2}\sigma^2 x^2\frac{\partial^2 F}{\partial x^2} + g\frac{\partial F}{\partial z} - rF = 0,$$
$$F(T, x, z) = \Phi(x, z).$$

Show further that the claim C_T can be hedged by a self-financing portfolio consisting at time t of

$$\phi_t = \frac{\partial F}{\partial x}(t, S_t, Z_t)$$

units of stock and

$$\psi_t = e^{-rt}\left(F(t, S_t, Z_t) - S_t\frac{\partial F}{\partial x}(t, S_t, Z_t)\right)$$

cash bonds.

13 Suppose that $\{N_t\}_{t\geq 0}$ is a Poisson process whose intensity under $\mathbb{P}$ is $\{\lambda_t\}_{t\geq 0}$. Show that $\{M_t\}_{t\geq 0}$ defined by

$$M_t = N_t - \int_0^t \lambda_s ds$$

is a $\mathbb{P}$-martingale with respect to the σ-field generated by $\{N_t\}_{t\geq 0}$.

14   Suppose that $\{N_t\}_{t\geq 0}$ is a Poisson process under $\mathbb{P}$ with intensity $\{\lambda_t\}_{t\geq 0}$ and $\{M_t\}_{t\geq 0}$ is the corresponding Poisson martingale. Check that for an $\{\mathcal{F}_t^M\}_{t\geq 0}$-predictable process $\{f_t\}_{t\geq 0}$,

$$\int_0^t f_s\,dM_s$$

is a $\mathbb{P}$-martingale.

15   Show that our analysis of §7.3 is still valid if we allow the coefficients in the stochastic differential equations driving the asset prices to be $\{\mathcal{F}_t\}_{t\geq 0}$-adapted processes, provided we make some boundedness assumptions that you should specify.

16   Show that the process $\{L_t\}_{t\geq 0}$ in Theorem 7.3.5 is the product of a Poisson exponential martingale and a Brownian exponential martingale and hence prove that it is a martingale.

17   Show that in the classical Black–Scholes model the *vega* for a European call (or put) option is strictly positive. Deduce that for vanilla options we can infer the volatility parameter of the Black–Scholes model from the price.

18   Suppose that $V(t, x)$ is the Black–Scholes price of a European call (or put) option at time t given that the stock price at time t is x. Prove that $\frac{\partial^2 V}{\partial x^2} \geq 0$.

19   Suppose that an asset price $\{S_t\}_{t\geq 0}$ follows a geometric Brownian motion with jumps occurring according to a Poisson process with constant intensity λ. At the time, τ, of each jump, independently, $S_\tau/S_{\tau-}$ has a lognormal distribution. Show that, for each fixed t, S_t has a lognormal distribution with the variance parameter σ^2 given by a multiple of a Poisson random variable.

Bibliography

Background reading:

- *Probability, an Introduction,* Geoffrey Grimmett and Dominic Welsh, Oxford University Press (1986).
- *Options, Futures and Other Derivative Securities,* John Hull, Prentice-Hall (Second edition 1993).

Grimmett & Welsh contains all the concepts that we assume from probability theory. Hull is popular with practitioners. It explains the operation of markets in some detail before turning to modelling.

Supplementary textbooks:

- *Arbitrage Theory in Continuous Time,* Tomas Björk, Oxford University Press (1998).
- *Dynamic Asset Pricing Theory,* Darrell Duffie, Princeton University Press (1992).
- *Introduction to Stochastic Calculus Applied to Finance,* Damien Lamberton and Bernard Lapeyre, translated by Nicolas Rabeau and François Mantion, Chapman and Hall (1996).
- *The Mathematics of Financial Derivatives,* Paul Wilmott, Sam Howison and Jeff Dewynne, Cambridge University Press (1995).

These all represent useful supplementary reading. The first three employ a variety of techniques while Wilmott, Howison & Dewynne is devoted exclusively to the partial differential equations approach.

Further topics in financial mathematics:

- *Financial Calculus: an Introduction to Derivatives Pricing,* Martin Baxter and Andrew Rennie, Cambridge University Press (1996).
- *Derivatives in Financial Markets with Stochastic Volatility,* Jean-Pierre Fouque, George Papanicolau and Ronnie Sircar, Cambridge University Press (2000).
- *Continuous Time Finance,* Robert Merton, Blackwell (1990).
- *Martingale Methods in Financial Modelling,* Marek Musiela and Marek Rutkowski, Springer-Verlag (1998).

Although aimed at practitioners rather than university courses, Chapter 5 of Baxter & Rennie provides a good starting point for the study of interest rates. Fouque, Papanicolau & Sircar is a highly accessible text that would provide an excellent basis for a special topic in a *second* course in financial mathematics. Merton is a synthesis of the remarkable research contributions of its Nobel-prize-winning author. Musiela & Rutkowski provides an encyclopaedic reference.

Brownian motion, martingales and stochastic calculus:

- *Introduction to Stochastic Integration,* Kai Lai Chung and Ruth Williams, Birkhäuser (Second edition 1990).
- *Stochastic Differential Equations and Diffusion Processes,* Nobuyuki Ikeda and Shinzo Watanabe, North-Holland (Second edition 1989).
- *Brownian Motion and Stochastic Calculus,* Ioannis Karatzas and Steven Shreve, Springer-Verlag (Second edition 1991).
- *Probability with Martingales,* David Williams, Cambridge University Press (1991).

Williams is an excellent introduction to discrete parameter martingales and much more (integration, conditional expectation, measure, ...). The others all deal with the continuous world. Chung & Williams is short enough that it can simply be read cover to cover.

A further useful reference is *Handbook of Brownian Motion: Facts and Formulae,* Andrei Borodin and Paavo Salminen, Birkhäuser (1996).

Additional references from the text:

- Louis Bachelier, La théorie de la speculation. *Ann Sci Ecole Norm Sup* **17** (1900), 21–86. English translation in *The Random Character of Stock Prices*, Paul Cootner (ed), MIT Press (1964), reprinted Risk Books (2000)
- J Cox, S Ross and M Rubinstein, Option pricing, a simplified approach. *J Financial Econ* **7** (1979), 229–63.
- M Davis, Mathematics of financial markets, in *Mathematics Unlimited – 2001 and Beyond,* Bjorn Engquist and Wilfried Schmid (eds), Springer-Verlag (2001).
- D Freedman, *Brownian Motion and Diffusion,* Holden-Day (1971).
- J M Harrison and D M Kreps, Martingales and arbitrage in multiperiod securities markets. *J Econ Theory* **20** (1979), 381–408
- J M Harrison and S R Pliska, Martingales and stochastic integrals in the theory of continuous trading. *Stoch Proc Appl* **11** (1981), 215–60.
- F B Knight, *Essentials of Brownian Motion and Diffusion,* Mathematical Surveys, volume 18, American Mathematical Society (1981).
- T J Lyons, Uncertain volatility and the risk-free synthesis of derivatives. *Appl Math Finance* **2** (1995), 117–33.
- P Protter, *Stochastic Integration and Differential Equations,* Springer-Verlag (1990).
- D Revuz and M Yor, *Continuous Martingales and Brownian Motion,* Springer-Verlag (Third edition 1998).
- P A Samuelson, Proof that properly anticipated prices fluctuate randomly, *Industrial Management Review* **6**, (1965), 41–50.

Notation

Financial instruments and the Black–Scholes model

T, maturity time.

C_T, value of claim at time T.

$\{S_n\}_{n \geq 0}$, $\{S_t\}_{t \geq 0}$, value of the underlying stock.

K, the strike price in a vanilla option.

$(S_T - K)_+ = \max\{(S_T - K), 0\}$.

r, continuously compounded interest rate.

σ, volatility.

$\mathbb{P}$, a probability measure, usually the market measure.

$\mathbb{Q}$, a martingale measure equivalent to the market measure.

$\mathbb{E}^{\mathbb{Q}}$, the expectation under $\mathbb{Q}$.

$\frac{d\mathbb{Q}}{d\mathbb{P}}$ the Radon–Nikodym derivative of $\mathbb{Q}$ with respect to $\mathbb{P}$.

$\{\tilde{S}_t\}_{t \geq 0}$, the *discounted* value of the underlying stock. In general, for a process $\{Y_t\}_{t \geq 0}$, $\tilde{Y}_t = Y_t / B_t$ where $\{B_t\}_{t \geq 0}$ is the value of the riskless cash bond at time t.

$V(t, x)$, the value of a portfolio at time t if the stock price $S_t = x$. Also the Black–Scholes price of an option.

General probability

$(\Omega, \mathcal{F}, \mathbb{P})$, probability triple.

$\mathbb{P}[A|B]$, conditional probability of A given B.

Φ, standard normal distribution function.

$p(t, x, y)$, transition density of Brownian motion.

$X \overset{\mathcal{D}}{=} Y$, the random variables X and Y have the same distribution.

$Z \sim N(0, 1)$, the random variable Z has a standard normal distribution.

$\mathbb{E}[X; A]$, see Definition 2.3.4.

Martingales and other stochastic processes

$\{M_t\}_{t \geq 0}$, a martingale under some specified probability measure.

$\{[M]_t\}_{t \geq 0}$, the quadratic variation of $\{M_t\}_{t \geq 0}$.

$\{\mathcal{F}_n\}_{n \geq 0}$, $\{\mathcal{F}_t\}_{t \geq 0}$, filtration.

$\{\mathcal{F}_n^X\}_{n\geq 0}$ (resp. $\{\mathcal{F}_t^X\}_{t\geq 0}$), filtration generated by the process $\{X_n\}_{n\geq 0}$ (resp. $\{X_t\}_{t\geq 0}$).

$\mathbb{E}[X\,|\,\mathcal{F}]$, $\mathbb{E}[X_{n+1}\,|\,X_n]$, conditional expectation; see pages 30ff.

$\{W_t\}_{t\geq 0}$, Brownian motion under a specified measure, usually the market measure.

$X^*(t)$, $X_*(t)$, maximum and minimum processes corresponding to $\{X_t\}_{t\geq 0}$.

Miscellaneous

$\overset{\Delta}{=}$, defined equal to.

$\delta(\pi)$, the mesh of the partition π.

$f|_x$, the function f evaluated at x.

θ^t (for a vector or matrix θ), the transpose of θ.

$x > 0$, $x \gg 0$ for a vector $x \in \mathbb{R}^n$, see page 11.

Index